Crop Production Systems Management

Editor
Dr. Ajoy Gangopadhyay

2007

Gene-Tech Books

New Delhi - 110 002

ISBN 81-89729-64-0

Published by : GENE-TECH BOOKS
4762-63/23, Ansari Road, Darya Ganj,
NEW DELHI - 110 002
Phone: 41562849
e-mail: genetechbooks@yahoo.co.in

Printed at : Tarun Offset Printers
Delhi

PRINTED IN INDIA

Preface

Crop production management deals with how farmers combine land, water, machinery, structures, commercial inputs, labour, and management skills to produce crop commodities. Management systems embody some of the more important decisions related to production, and include nutrient management, soil management, water management, weed management, and the like. The overall production management system can be thought of as the combination of activities chosen for each aspect of crop production.

Production decisions concerning what crops to produce, how much effort and resources to invest, and which physical practices to follow have immediate consequences for the farm and long-term effects on the environment. This book examines the management of farm businesses; and the conventional and organic production systems employed, including soil, nutrient, pest, and irrigation water management. Data pertaining to Indian conditions has been given to help the reader gain an insight into India's own crop production problems. The contents of the book will be of immense value to students of agriculture, researchers and decision makers associated with crop production management, soil and water pollution, sewage water irrigation, forestry and environment management.

Editor

Contents

1

Cropping Patterns

Cropping activities go on all the year-round in India, provided water is available for crops. In northern India, there are two distinct seasons, kharif (July to October), and rabi (October to March). Crops grown between March and June are known as zaid. In some parts of the country, there are no such distinct seasons, but there they have their own classification of seasons. The village revenue officials keep plot-wise record of crops grown in each season. These are annually compiled district-wise, state-wise and on all-India basis.

From these records one could calculate the relative abundance of a crop or a group of crops in a region. These crops are grown sole or mixed (mixed-cropping), or in a definite sequence (rotational cropping). The land may be occupied by one crop during one season (mono-cropping), or by two crops (double-cropping) which may be grown in a year in sequence. Of late, the trend is even more than two crops (multiple-cropping) in a year. These intensive croppings may be done either in sequence or even there may be relay-cropping-one crop undersown in a standing crop.

With wide-rowed slow growing cropping patterns, companion crops may be grown. There are various ways of utilising the land intensively. It is proposed to give a synoptic view of cropping patterns prevalent in the country. Before dealing with the cropping patterns, a brief description of the factors that determine the cropping systems of an individual locality or region are briefly presented here.

In any locality, the prevalent cropping systems are the cumulative results of past and present decisions by individuals, communities or governments and their agencies. These decisions are usually based on experience, tradition, expected profit, personal preferences and resources, social and political pressures and so on. Essentially, they are answers to some of the following questions:

— What with the present pest-and-disease control methods are ecologically practicable?

— What interactions occur among the ecologically practicable crops, and the chosen crops and must be combined in a special way (rotations) in the farming systems?

— Are any of the ecologically feasible crops ruled out by infrastructural factors?

— Which of the crops, now remaining on the list, are most profitable (or yield most food in a subsistence agriculture)?. In what combinations and at what level of input application would they make the best use of local land, climate and input resources in short-term and long-term situations bearing in mind the degree of food and income security required by the individual farmer and the community?

— What operational factors rule out or amend the size and the method of any of the economically preferable crop combinations thereof?

— Finally, are the crop combinations, the farming systems and the input levels suggested by this process of the individual farmers compatible with his own skills, enterprise preferences, health, age and capital?

The climatic, edaphic and socio-economic diversity of the Indian crop-production scene is dotted with many cropping patterns. With a geographic area of 328.048 million hectares, stretching between 8°N and 36°N latitude and between 68°E and 98°E longitude, its altitude varying from the mean sea-level to the highest mountain ranges of the world, India presents a range and diversity of climate, flora and fauna, with a few parallels in the world.

The country presents a paradox of containing in it the station with the highest mean annual rainfall in the world (Cherrapunji in Assam) and also dry, semi-desert area in Rajasthan. The variability of rainfall is most important in all the states, but especially where rainfall is low. In parts of Rajasthan and the Deccan, the variability is more than 100 per cent of the mean. Years of drought account for only too frequent a history of crop failures, whereas the years of flood also cause very considerable loss of agricultural production.

Temperatures also vary greatly, both geographically and seasonally. Northern and central parts of India in January have temperature comparable with those in Europe in July, though with a greater daily range, but in these places in the pre-monsoon months the maximum temperatures of over 40°C are reached over a large area. Frost may occur in winter in the plains, as far south as a line drawn through Madhya Pradesh and may be heavy in Kashmir and areas north of Punjab.

Socio-economically, the peasantry ranges from the relatively affluent Punjabi farmers who operate with a high input intensity in agriculture to the subsistent farmers of eastern and central India. They even today, sometimes practice shifting cultivation. Between these two extremes, various intensities of cultivation are practised. The outstanding fact on the socio-economic is the smallness of holdings, the average farm-size in most areas being lower than that is in most tropical countries.

Crops production, therefore, presents such an enormous diversity owing to differences in latitude, altitude and variability of rainfall and edaphic diversity which have presented in detail in the book. Thus it may not be possible to enumerate and describe here every type of cropping pattern prevalent in the country. Some broad contours of farming, however, emerge. The most important element of farming in India is the production of grains and the dominant food-chain is grain-man. On this basis, the country may be divided broadly into five agricultural regions.

— The rice region extending from the eastern part to include a very large part of the north-eastern and the south-eastern India, with another strip along the western coast.

— The wheat region, occupying most of the northern, western and central India.

— The millet-sorghum region, comprising Rajasthan, Madhya Pradesh and the Deccan Plateau in the centre of the Indian Peninsula.

— The temperate Himalayan region of Kashmir, Himachal Pradesh and Uttar Pradesh and some adjoining areas. Here potatoes are as important as cereal crops, and the tree-fruits form a large part of agricultural production.

— The plantation crops region of Assam and the hills of southern India where good quality tea is produced. There is an important production of high-quality coffee in the hills of the western peninsular India. Rubber is mostly grown in Kerala and parts of Karnataka and Tamil Nadu. There are some large estates, but most of the growers would come under the category of small holders. Sugarcane, which in many countries is a plantation crop, is almost entirely grown by small holders in India.

There had been substantial investments in major irrigation works in the colonial days. The post-Independence era saw many multi-purpose irrigation works. Lately, interest in the medium and minor irigation works has increased, especially after the drought of 1966. Thus, at present, an all-India irrigation potential of 38.5m ha has been created and is expected to increase up 110 m ha by 2025. Irrigation, especially the minor works, has provided a base for multiple-cropping.

The All-India Co-ordinated Crops-Improvement Projects run co-operatively by the Indian Council of Agricultural Research and the agricultural universities have generated short-season, photo-period-insensitive high-yeilding varieties of various crops suitable for a high intensity of cropping. The adaptability of these varieties on the farmer's fields has been demonstrated in the National Demonstration Programme spread all over the country.

The various developmental and the educative programmes, especially the High Yielding Varieties Programme, have also resulted in newer cropping patterns involving intensive cropping. The area of rice has increased in Punjab and Haryana. Similarly wheat is now grown in West Bengal and to some extent in the southern states of the country.

All these factors have led to the present cropping patterns, which are getting more and more intensive both in respect of the number of crops grown per year and in respect of the intensity of inputs utilised in the production of these crops.

A broad picture of the major cropping patterns in India can be presented by taking the major crops into consideration. To begin with, the south-westerly monsoon crops (kharif), bajra, maize, ragi, groundnut and cotton. Among the post-monsoon crops (rabi), wheat, sorghum (rabi) and gram can also be considered to be the base crops for describing the cropping patterns. With such an approach, the crop occupying the highest percentage of the sown area of the region is taken as the base crop and all other possible alternative crops which are sown in the region either as substitutes of the base crop in the same season or as the crops which fit in the rotation in the subsequent season, are considered in the pattern.

Also these crops have been identified as associating themselves with a particular type of agroclimate, and certain other minor crops with similar requirements are grouped in one category. For example, wheat, barley and oats, are taken as one category. Similarly the minor millets (Paspalum, Setaria and Panicum spp.) are grouped with sorghum or bajra. Certain other crops, such as the plantation crops and other industrial crops are discussed separately.

Kharif Cropping Patterns

Among the kharif crops, rice, jowar, bajra, maize, groundnut and cotton are the prominent crops to be considered the base crops for describing the kharif cropping patterns.

Rice-based Cropping Patterns

Rice is grown in the high-rainfall area or in areas where supplemental irrigation is available to ensure good yields. If the crop has to depend solely on rainfall, it requires not less than 30 cm per month of rainfall over the entire growing period. However, only 9 per cent of the area in the country comes under this category, and it lies in the eastern parts. Nearly 45 per cent of the total rice area in India receives 30 cm per month of rainfall during at least two months (July and August) of the south-westerly monsoon and much less during other months.

In contrast to these parts, the eastern and southern regions comprising Assam, West Bengal, coastal Orissa, coastal Andhra Pradesh, Karnataka (most part), Tamil nadu and Kerala receive rainfall of 10 to 20 cm per month in four to eight consecutive months, starting earlier or going over later than the south-westerly monsoon months. With supplemental irrigation, 2 or 3 crops are taken in these areas. However, it has been observed that on an all-India basis, nearly 80% of rice is sown during June-September and the rest during the rest of the season. Areawise the mono-season belt occupies 53.6 per cent of the area (comprising Assam, West Bengal, coastal Orissa, coastal Andhra Pradesh, parts of Tamil Nadu, Karnataka and Kerala).

On an all-India basis, about 30 rice-based cropping patterns have been identified in different states. In the most humid areas of eastern India comprising Tripura, Manipur and Mizoram, rice is the exclusive crop. In Meghalaya, rice is alternated with cotton, vegetable and food-crops, whereas in Arunachal Pradesh, where rice is not grown exclusively, the alternative crops being maize, small millets and oilseeds. In parts of Assam, West Bengal, Bihar, orissa and northern coastal districts of Andhra Pradesh, jute forms an important commercial crop alternative to rice.

In West Bengal, besides rice and jute, pulses and maize are grown on a limited scale. In Bihar, rice is grown over 49 per cent (5.3 m ha) of its cropped area (14.2 per cent of all-India area), whereas pulses, wheat, jute, maize, sugarcane and oilseeds are the alternative crops. In Uttar Pradesh rice is grown on 19 per cent (4.6 m ha) of its cropped area and represents about 12.4 per cent of the all-India area under this crop. Rice is concentrated in the eastern districts of Uttar Pradesh where the alternative crops are pulses, groundnut, sugarcane, bajra and jowar in the decreasing order of their importance. Tobacco is grown in some districts.

In Orissa, rice is grown on more than 50 per cent of the area, whereas the alternative crops are: pulses, ragi, oilseeds, maize and small millets. In Madhya Pradesh rice is grown in the Chattisgarh area on 4.3 m ha (11.7 per cent of the all-India rice area), but the crop suffers because of inadequate rainfall and irrigation. The important alternative crops of this area are: small millets, pulses and groundnut. Wheat is also grown on a limited scale.

In the southern states, namely Andhra Pradesh, Tamil Nadu and Kerala rice is grown in more than one season and mostly under irrigation or under sufficient rainfall. Together, these three states have over 6.0 m ha, representing over 17 per cent of the all-India area under rice. Important alternative plantation crops in Andhra Pradesh are: pulses, groundnut, jowar, maize, sugarcane and tobacco. In Karnataka the crops alternative to rice are: ragi, plantation crops, bajra, cotton, groundnut, jowar and maize. In Kerala plantation crops and tapioca form the main plantation crops alternative to rice.

In Maharashtra rice is grown mostly in the Konkan area over 1.3 m ha, along with ragi, pulses, rabi jowar, sugarcane, groundnuts and oilseeds. In other states, namely Gujarat, Jammu and Kashmir, Rajasthan and Himachal Pradesh, rice forms a minor plantation crop and is mostly grown with irrigation. However, in Punjab and Haryana and to some extent in western Uttar Pradesh owing to high water-table during this monsoon season, rice has become a major crops in such areas.

Kharif Cereals other than Rice

Maize, jowar and bajra form the main kharif cereals, whereas ragi and small millets come next and are grown on a limited area. By and large, maize is a crop grown commonly in high-rainfall areas, or on soils with a better capacity for retaining moisture, but with good drainage. Next comes jowar in the medium rainfall regions whereas bajra has been the main crop in areas with low or less dependable rainfall and on light textured soils.

The extent of the area under these crops during the south-westerly monsoon season is maize, 5.6 m ha; jowar (kharif), 11 mha, and bajra,12.4 m ha. Even though these crops are spread all over the western, northern and southern India, the regions of these crops patterns are demarcated well to the west of 80° longitude (except that of maize). Ragi as a kharif cereal is mainly concentrated in Karnataka, Tamil Nadu and Andhra Pradesh which account for main than 60 per cent of the total area under this crop in India. The cropping patterns based on each of these kharif cereals are discussed.

Maize-based Cropping Patterns

The largest area under the kharif maize is in Uttar Pradesh (1.4m ha), followed by Bihar (0.96 m ha), Rajasthan (0.78 m ha), Madhya Pradesh (0.58 m ha) and Punjab (0.52 m ha). In four states namely Gujarat, Jammu and Kashmir, Himachal Pradesh and Andhra Pradesh, the area under maize ranges from 0.24 to 0.28 m ha in each, whereas other states have much less area under it. Taking the rainfall of the maize growing areas under consideration, over 72 per cent of the areas receive 20-30 cm per month of rainfall in at least two months or more during the south westerly monsoon season.

On the all-India basis, about 12 cropping patterns have been identified. They have maize as the base crop. In the maize growing areas of Uttar Pradesh and Bihar, rice in kharif and wheat in rabi are the main alternative crops. In some areas, bajra, groundnut, sugarcane, ragi and pulses are taken as alternative crops. In Rajasthan maize is grown as an extensive crop in some areas, whereas at other places, it is replaced by small millets, pulses, groundnut and wheat(rabi) as alternative crop. In madhya Pradesh mainly the kharif jowar is replaced by maize, whereas rice and groundnut are also grown to a limited extent. In Punjab maize has groundnut, fodder crops and wheat(rabi) as alternative crops. In other states, e.g. Gujarat, rice, groundnut, cotton and wheat form the alternative crops in the maize-growing areas of Himachal Pradesh, whereas in Andhra Pradesh, rice, kharif jowar, and oilseeds are grown in these areas.

Kharif Jowar-based Cropping Patterns

The area under the kharif jowar in India is highest in Maharashtra (2.5 m ha), closely followed by madhya Pradesh (2.3 m ha), whereas in each of the states of

Rajasthan, Andhra Pradesh, Karnataka and Gujarat, the area under this crops is between 1.0 and 1.4 m ha. Jowar is mainly grown where rainfall distribution ranges from 10-20 per month at least for 3 to 4 months of the south-westerly monsoon or is still more abundant. On the all-India basis, about 17 major cropping patterns have been identified. In them the base crops is kharif jowar. Most of the alternative crops are also of the type which can be grown under medium rainfall.

In Maharashtra cotton, pulses, groundnut and small millets are sown as alternative crops. In the adjacent states of Madhya Pradesh, besides the above alternative crops, wheat and fodder are sown. In Rajasthan wheat, cotton, bajra and maize are grown in the kharif-jowar tract, whereas in Andhra Pradesh, groundnuts, cotton, oilseeds and pulses form the main alternative crops. Besides cotton and groundnut, ragi is sown in the kharif-jowar tract of Karnataka, whereas in Gujarat, bajra, cotton and groundnut are the major alternative crops.

Bajra-based Cropping Patterns

Bajra is more drought-resistant crop than several other cereal crops and is generally preferred in low-rainfall areas and on light soils. The area under the bajra crop in India is about 12.4 m ha and Rajasthan (4.6 m ha) shares about the 2/3 total area. Maharashtra, Gujarat and Uttar Pradesh together have over 4.6 m ha, constituting an additional 1/3 area under bajra, in India. Over 66 per cent of this crop is grown in areas receiving 10-20 cm per month of rainfall, extending over 1 to 4 months of the south-westerly monsoon.

On the all-India basis, about 20 major cropping patterns have been identified with bajra. However, it may be observed that jowar and bajra are grown mostly under identical environmental conditions and both have a wide spectrum adaptability in respect of rainfall, temperature and rainfall.

Considering the cropping patterns in different states, bajra is grown along with pulses, groundnut, oilseeds and kharif jowar in Rajasthan. Gujarat has a similar cropping pattern in its bajra areas, except that cotton and tobacco are also grown. In Maharashtra besides having some areas solely under bajra, pulses, wheat, rabi jowar, groundnut and cotton are substituted for it. In Uttar Pradesh, maize, rice and wheat form the main alternative crops to this crop.

Groundnut-based Cropping Patterns

Groundnut is sown over an area of about 7.2 m ha, mostly in five major groundnut-producing states of Gujarat (24.4 per cent area), Andhra Pradesh (20.2) per cent), Tamil Nadu (13.5 per cent), Maharashtra (12.2 per cent) and Karnataka (12.0 per cent). Five other states viz. Madhya Pradesh, Uttar Pradesh, Punjab, Rajasthan and Orissa together have about 17.3 per cent of the total area under this crop.

The rainfall in the groundnut area ranges from 20-30 cm per month in one of the monsoon months and much less in the other months. In some cases the rainfall is even less than 10 cm. per month during the growth of the crop. The irrigated area under groundnut is very small and that too, in a few states only, viz. Punjab (16.4 per cent), Tamil Nadu (13.3 per cent) and Andhra Pradesh (12.5 per cent).

On the all-India level, about 9 cropping patterns have been identified with this crop. In Gujarat besides the sole crop of groundnut in some areas, bajra, is the major alternative crop, whereas the kharif jowar, cotton and pulses are also grown in this tract. In Andhra Pradesh and Tamil Nadu, this crop receives irrigation in some areas and rice forms an alternative crop. Under rainfed conditions, bajra, kharif jowar, small millets, cotton and pulses are grown as alternative crops. In Maharashtra both the kharif and rabi jowar and small millets are important alternative crops. In Karnataka also, jowar is the major alternative crop, whereas cotton, tobacco, sugarcane and wheat are also grown in this tract.

Cotton-based Cropping Patterns

Cotton is grown over 7.6 m ha in India. Maharashtra shares 36 per cent (2.8 m ha), followed by Gujarat with 21 per cent (1.6 m ha), Karnataka with 13 per cent (1 m ha) and Madhya Pradesh with 9 per cent (0.6 m ha) of the area. Together, these four states account for about 80 per cent of the area under cotton. Other cotton-growing states with smaller areas are Punjab, with 5 per cent (0.4 m ha), Andhra Pradesh and Tamil Nadu each with 4 per cent (0.31 m ha), Haryana and Rajasthan with 3 per cent of each (0.2 m ha each).

Most of the cotton areas in the country are under the high to medium rainfall zone. The cotton grown in Madhya Pradesh, Maharashtra, Karnataka, and Andhra Pradesh (4.8 m ha) is rainfed, whereas in Gujarat and Tamil Nadu (1.93 m ha) it receives partial irrigation 16-20 per cent of the area). The area under cotton in Punjab, Haryana, Rajasthan and Uttar Pradesh (0.8 m ha) gets adequate irigation, ranging from 71 to 97 per cent of the area. These growing conditions, together with the species of cotton grown, determine the duration of the crop which may vary from about 5 to 9 months.

On the all-India basis, about 16 broad cropping patterns have been identified. In Maharashtra, Madhya Pradesh, Andhra Pradesh and Karnataka, the cropping patterns in the cotton-growing areas are mostly similar owing to identical rainfall. These patterns include jowar (kharif and rabi), groundnut and small millets. Pulses and wheat are also grown in a limited area. In some pockets, when irrigation is available, rice and sugarcane are also grown. In Gujarat, rice, tobacco and maize are grown, besides the rainfed crops, e.g. jowar and bajra.

Rabi Season Cropping Patterns

Among the rabi crops, wheat, together with barley and oats, jowar and gram, are the main base crops among the rabi cropping patterns. Generally, wheat and gram are concentrated in the subtropical region in northern India, whereas the rabi sorghum is grown mostly in the Deccan. The extent of these areas in different states is as follows;

Crop	Area	Region (per cent of all-India area)
Sugarcane	2.5 m ha	Uttar Pradesh (51), Haryana (6),Bihar (6), Punjab (6), Maharashtra (8), Andhra Pradesh (5),Tamil Nadu (5), Karnataka (3)
Tobacco	0.427 m ha	Andhra Pradesh (48), Gujarat (19.5), Karnataka (8.7), Maharashtra (3.5), Tamil Nadu (3.5)
Potato	0.491	Uttar Pradesh (33.6), Bihar (20.4), West Bengal (13.3), Assam (5.2), Orissa (4.8)
Jute	0.778	West Bengal (60), North eastern Region (18.7), Bihar (17.6), Orissa (6.1), Uttar Pradesh (1.7)
Coconut	1.05 m ha	Kerala (68.3), Karnataka (12.4), Tamil Nadu (9.7), Andhra Pradesh (3.5)
Rubber	0.197 m ha	Kerala (92.8), Tamil Nadu (5.0), Karnataka (1.9)
Cashew	0.264 m ha	Kerala (67.4), Karnataka (12.1), Andhra Pradesh (10.8), Tamil Nadu (9.8), Maharashtra (4.8)
Tea	0.35 m ha	West Bengal, Assam and Tripura (77), Kerala, Tamil Nadu and Karnataka (20)

Coffee	0.138 m ha	Kerala , Tamil Nadu and Karnataka (99)
All fruit-crops"		1.8 m haSpread all-over India
Onion	0.16 m ha	Maharashtra (18.5), Karnataka (11.7), Andhra Pradesh (12.8), Tamil Nadu (11.2), West Bengal (7.6), Madhya Pradesh (7.2), Orissa (6.8), Punjab (6.2)
Chillies	0.733 m ha	Andhra Pradesh (26.9), Maharashtra (20.4), Karnataka(14.5), Madhya Pradesh (5.5), Tamil Nadu (10.1)
Coriander	0.283 m ha	Andhra Pradesh (36), Rajasthan (23.6), Madhya Pradesh (11.1), Tamil Nadu (10.0)

In several sugarcane-growing areas, mono-cropping is practised, and during the interval between the crops, short duration seasonal crops are grown. In U.P., Bihar, Punjab and Haryana, wheat and maize are the rotation crops. rice is also grown in some areas. In the southern states, namely Tamil Nadu, Karnataka and Andhra Pradesh, ragi, rice and pulses are grown along with sugarcane. In Maharashtra, pulses, jowar and cotton are grown.

In the potato-growing region, maize, pulses, wheat are the alternative crops. In the tobacco-growing areas, depending on the season and the type of tobacco, jowar, oilseeeds and maize are grown in rotation. In the jute-growing areas, rice is the usual alternative crop. In the case of plantation-crops, intercropping with pulses and fodder crops is common. Spices and condiments are generally grown on fertile soils. Chillies are rotated with jowar, whereas onion, corriander, turmeric and ginger are grown as mixed crops with other seasonal crops.

Mixed Cropping

Crops mixtures are widely grown, especially during the kharif season. Pulses and some oilseeds are grown with maize, jowar and bajra. Lowland rice is invariably grown unmixed, but in the case of upland rice, several mixtures are prevalent in eastern Uttar Pradesh, with Chotanagpur Division of Bihar and in the Chhatisgarh Division of Madhya Pradesh. During the rabi season, especially in the unirrigated area of the north, wheat and barley and wheat and gram or wheat + barley + gram are the mixtures of grain crops. Brassica and safflower are grown mixed with gram or even with wheat.

Mixed cropping was considered by researchers a primitive practice, but now many researchers regard mixed cropping as the most efficient way of using land. Several new mixtures have recently been suggested. They ensure an efficient utilisation of sunshine and land. Breeders are developing plant types in pulses and oilseeds, with good compatibility with row crops.

Future of Cropping Patterns

With the increase in population, the irrigated area is increasing and with advances in agricultural science, most of the extensive cropping patterns are giving way to intensive cropping. The development in minor irrigation works has especially provided the farmes with opportunities to crop their land all the year round with high-yielding varieties. This intensive cropping will require an easy and ready availability of balanced fertilisers and plant protection chemicals and an appropriate price policy for inputs and agricultural produce. India is a country of small farmers. In the future the size of the holdings will diminish further. The country has to produce enough for its people without deteriorating the quality of the environment. This is the challenge of the future for the farmers, agricultural scientists, extension workers and administrators.

2

Farm Management

A farm is the basic unit in agriculture. It is a section of land devoted to the production and management of food, either produce or livestock. It may be an enterprise owned and operated by a single individual, family, or community, or it may be owned by a corporation or company. A farm can be a holding of any size from a fraction of a hectare to several thousand hectares.

The word farm has its roots in the Anglo-Saxon word feorm, which relates to provisioning and food supply, and was originally indicative of a form of taxation, whereby goods or monetary equivalents were liable to the king. Over time, this taxation was translated into a form of rental tax.

The development of farming and farms was an important component in establishing towns. Once people have moved from hunting and/or collecting and from simple horticulture to active farming, social arrangements of roads, distribution, collection, and marketing can evolve. With the exception of plantations and colonial farms, farm sizes tend to be small in newly-settled lands and to extend as transportation and markets become sophisticated. Farming rights have been the central tenet of a number of revolutions, wars of liberation, and post-colonial economics.

Farm management is different from what is commonly confused with the work of a farm manager who manages a government farm as an agronomist. His function is normally limited to supervising and handling the day-to-day routine of a farm. It normally pertains to the existing pattern of the resource-use and crops-mix under which only the existing plan is executed, supervised and carried out. An intelligent farm manager may go a little further and look after the farm machinery to keep it going.

Farm management, is however, much more than that. Here we are not just concerned with the distribution of labour and irrigation water for day-to-day operations. The emphasis is on the decision-making function of evaluating and choosing between alternative strategies. A major concern is about adjustments which are more suitable and profitable and about exploring new situations and opportunities for maximisation of income and satisfying other goals of a farmer. It is the approach under which the opportunity costs of the various resources are evaluated and adjustments in resource-use and enterprise mix are made to secure higher levels of farm income.

Following this approach, the emphasis on yield (productivity per unit of a resource, mainly land) is not ignored, but the greater focus is on the increasing of farm income through a second business organisation. As a business, farming requires the application of business methods and efficient management. To be an efficient manager, one must keep oneself abreast of developments in new technology, new practises, price trends, and economic outlook. Again a farm management man should identify the constraints in the external environment conditions, which hamper a farmer's opportunities and plans for making adjustments in his farm organisation and render the technically superior production plan economically unattractive to the farmer.

Basically farm management is economics in the context of fundamental definition of economics which involves three elements, viz. the scarcity of resources. Their alternative uses and the objective of profit maximisation. It is this science which deals with making rational, profitable and economic recommendations to the farmers. It also deals with the growth and stability aspects of economics.

The basic information about the multiple (alternative) uses to which the scarce resources can be allocated is supplied by other physical and biological sciences. The research in these sciences should continually generate the relevant data on alternative technologies and practices whose profitability can be tested under actual farm situations. And these data are required for the farm organisation as a whole and not far for a single hectare, animal or tree.

The challenge to the farm management specialist is to be able to integrate and synthesise the diverse pieces of information from many disciplines such as agronomy, animal husbandry, horticulture, soil science, plant breeding, entomology, plant pathology, general economics, sociology and psychology, into an optimum 'package' which can be used on the farms with profit.

Agricultural scientists mainly put emphasis on the maximisation of yield rather than on the use of the optimum level of resources. But the goal in farming is not to make a profit on some single enterprise or from a part of the farm land as some krishi pandits do, but to use land, labour, and capital resources in such a way that they make the greatest contribution to the total profits from the entire farm. The superiority of this discipline, thus, lies in its treating the farm as an operational unit and tailoring the recommendations of all other disciplines to fit into an individual farmer's pattern of resources.

In the context of the recent technological breakthrough, management today should be viewed as a process within a rapidly moving frame of reference. " It is now more scientific, less artistic; more dynamic, less static; more versatile and less rigid". Farm management is forward-looking in its approach. Its task is not so much the improvement of the present farming practices but of the establishment of the whole sets of new production methods and farming systems which would put our agriculture on a continuously rising growth curve.

Government policies should change the economic environment to help the interests of the farmers to converge on the national goals. Studies on farm management to determine the responsiveness of the farmers to different levels of prices become extremely relevant in inducing the farmers to produce the quantities of different agricultural products and services needed by society. A rational pricing policy for water is needed. Canal water, for e.g. is charged for without any direct relation to the quantity supplied.

As a result, there is no incentive to the farmer to allocate water in the most economical manner. Farm management helps to identify such uneconomic practices and the most limiting factors. Irrigation or power (the bullock versus the tractor) may be a more important restriction than the limits imposed by land. In areas where water rather than land is the principal limiting factor and the marginal-value productivity of irrigation water is very high, it should be most carefully used rather than over irrigating a few hectares of crops needing high water consumption.

Good farm management can lead to a highly productive use of farm resources and can avail itself of the technological revolution now going on in our agriculture.

Benefits of Farm Management

As against the general contention that farm management is mostly applicable to crop farms, its application to other types of farming business, such as orchards, dairying, poultry, etc. is no less important. Faculty farm management has contributed more to citrus decline than the poor physico-chemical conditions of the soil, the stock-scion incompatibility, insect pests, disease and viruses.

Intercropping with exhaustive crops seems to be one of the major reasons for the poor health of the plants. Citrus die-back or decline does not seem to be a disease by itself, but a mere symptomatic expression of physiological imbalance owing to the cumulative effect of several factors. Similarly, if milk production has to be encouraged, it has to be made more profitable than cereal production. One has to demonstrate how the addition of milch animals to a farm fits into its resource organisation and yields higher returns.

With land as a scarce resource, efforts are made to intensify its use for maximising profits through multiple cropping. During a given period, the relative profitability of different crop rotations is evaluated and finally selected for adoption. 'Kufri Sinduri' variety of potato is not being adopted on a large scale in spite of its giving 10 percent higher yield than some other varieties, e.g. 'Up-to-date'. The reason is simple. The former is a late variety and delays the sowing of wheat which decreases its yield. It is not 'Kufri Sinduri' potatoes versus 'Up-to-date' potatoes; rather, it is 'Kufri-Sinduri' followed by wheat versus 'Up-to-date' followed by wheat that counts in farm management.

Similarly, in West Bengal, where jute has been competing with paddy, the situation is changing with the introduction of high-yielding varieties of wheat, which can be rotated with paddy. The comparison in terms of the relative profitability now runs between jute and paddy plus wheat, and not just between jute and paddy. Owing to the paddy-wheat rotation becoming more profitable, jute production in West Bengal is suffering a set back. As Indian agriculture becomes more productive with the passage of time, the farmers will be confronted with the task of making the best selection from among many alternatives.

Sometimes it is said that management is not important on small farms. This notion is widely prevalent under those situations where the farms operate on a low level of technology. In the case of small farms, farm organisation cannot be changed as much as on a large farm, but the choice in respect of farm practices and methods of production , cropping intensity, etc. offer worthwile alternatives. Japan's method of improving agriculture based on the principles of farm management clearly demonstrates that a small farm by itself should not be a hindrance to increased production and higher income.

Small farms may have different problems, but the principles of the efficient use of available resources to obtain the maximum economic returns and family satisfaction remain the same in both the cases. The principles of farm management, since they deal with the allocation of resources, apply to small and large farms equally.

Farm Planning

Planning means taking decisions in advance. It stimulates thinking, broadens understanding and challenges the farmer to move forward. It is a forward-looking approach. The farm plan helps a farmer to decide how to combine new ideas and old ones to his best advantage. By identifying his credit and supply needs, the farm plan helps him to arrange for the timely supplies of credit, seeds, fertilisers, etc. A specific farm plan setting fort his expected output, expenses and income, serves as a sound basis on which a credit institution can give him production credit, based on his productive capability rather than on his net financial assets.

It is out of his income and not through the sale of assets that the cultivator has to pay off his loan. Thus the farm plan or the budget is to the farmer what the blue-print of the architect is to a building contractor. It shows what is to be done and how it is to be done. It furnishes an organised and logical approach to his problems and helps him to work out the solution.

Types of Farm Budgeting

There are three main types of farm budgeting:

— Enterprise budgeting

— Partial budgeting

— Full or complete budgeting and planning.

Enterprise Budgeting

The enterprise budgets are the input-output relationship for individual enterprises. An enterprise budget includes all the variable resources required per unit (a hectare/ animal/tree, etc.) of an enterprise and its cost, the expected output, gross returns, etc. A format of an enterprise budget is given in Table 1.

Table 1: Format of an enterprise budget

Items	Quantity (kg/litre)	Value (Rs)
GROSS RETURNS		
Main product		
By-product		
Total		
CASH VARIABLE EXPENSES		
1. Seed and seed treatment		
(i)Seed		
(ii)Sub-total		
2. Manures and fertilisers		
A. Farmyard manure		
B. Chemical fertilisers		
(i) CAN urea		
(ii) Superphosphate		
(iii) Muriate of Potash		
Sub-total		
3. Insecticides and fungicides		
4. Tractor fuel cost		
5. Irrigation hours		
6. Human labour		
7. Threshing hours with diesel engine		
8. Cost of typing material		
9. Marketing charges		
Sub-total		
10. Interest on variable expenses for half the period of growth		
11. Total variable expenses		
12. Returns over variable expenses		
13. Man-hours		
14. Bullock labour (pair hours)		
15. Machine (tractor hours)		

Enterprise budgets provide useful information regarding the resources requirements and the relative profitability of different enterprises. Thus these budgets, considered in the framework of farm resources, are the alternatives from among which the most profitable ones are to be selected. In this context, the enterprise budgets need to be prepared at different levels of technology, as:

(a) the existing level of technology; and

(b) the improved or recommended level of technology.

A comparison of the enterprise budgets at the existing and improved levels of technology provides the scope or the potential of making farm improvements. The enterprise budgets lack in one important aspect that these do not consider the complementary and supplementary relationships amongst themselves which are quite common among farm enterprises at low level of production, but they simply assume to be competitive to one another right from the beginning. But these relationships are taken care of in complete planing and budgeting.

Partial Budgeting

Partial budgeting is a method of making a comparative study of the cost-and-return analysis resulting from a change in a part of the business organisation. This change may be made through a careful selection from among alternative methods of production or practices, the choice of which is based on the opportunity cost of relative profitability and does not affect the total farm organisation vitally. This technique helps to make decisions whenever small changes in the existing farm organisations are contemplated. The following four points are important in setting up a partial budget:

—Additional returns from change

—Reduction in unit cost

—Reduction in yield, if any

—Addition in cost incurred

Put in a format, a partial budget will look like this:

Debts	Rs	Credit	Rs
(a) Increase in costs		(a) Decrease in costs	
(b) Decrease in returns		(b) Increase in returns	
Gain		Loss	
Total		Total	

Thus partial budgets deal with such changes in the farm organisation as can increase farm incomes without changing the total farm organisation. The farmer would know the total net benefit from the change, the complete details of what he should do at what cost and what he is not to do after the change and come out with higher profits.

Partial budgeting maybe used where the change in the activity under study would not affect the farm organisation vitally. It has wide application and can be used to answer a wide variety of questions such as :

- — The economic combination of CAN and superphosphate versus diammonium phosphate with a basal dose of CAN,
- — the mechanical thinning of fruits like plums and grapes versus chemical thinning,
- — the substituting of a more economical dairy ration for a costlier one (a single or two nutrients), and
- — American cotton versus hybrid maize versus desi cotton, etc.

Partial budgeting considers a few alternatives which do not affect the organisation vitally, but full budgeting takes care of all the alternatives.

In full budgeting, the inventory of the farm, the resource structure, the existing resource use and such problems as overstocking or understocking of resources, etc. are considered, but in partial budgeting information with respect to a few alternatives is considered.

Partial budgeting does not indicate the break-even point as to when to start one practice and abandon another, but full budgeting does. The best strategy is to make an effective use of the partial budget at different stages of full budgeting.

The partial budgeting analysis suffers from some limitations, viz. it does not consider all the alternatives open to a farmer within the restraints of his present resources. It only considers a few alternatives and these too, should not affect the farm organisation to a great extent.

Partial budgeting does not always provide a complete solution. As one may demonstrate that mechanical threshing costs less per quintal than the traditional phalla method, but this would not serve the purpose. The other questions such as the break-even point of the thresher, i.e. the minimum volume of business (i.e. 1,000 quintals of wheat to thresh) the possibilities of custom work, the capital available to purchase the thresher, etc. should also be considered.

Similarly, the question whether to increase the number of dairy animals may be answered by using partial budgeting. But, in this case the partial budgeting analysis will approach very closely to a change in the complete farm organisation. Again, the partial budget may show American cotton to be less profitable than hybrid maize, but when full budget is prepared, both the activities may not enter the farm organisation.

Like any other farm-management analyasis, the result of partial budgets are subject to variations in the input-output prices and field conditions such as soil type, soil fertility, etc. Also the restraints pertinent to a particular farm organisation necessitate the tailoring of these results to ensure their applicability to means for making profits.

Complete Planning and Budgeting

In preparing complete plans for the farm, all the sophisticated analysis of studying an individual cultivator's opportunities, constraints and problems is done. Every little aspect of the farm organisation is examined and then suitable adjustments are studied and a suitable combination as fits in rationally with that given farm organisation is suggested. Various recommendations in terms of augmenting resources, if the need may arise, are also made. It implies the following steps.

1. *Farm map*. The farm is carefully mapped out, giving its salient features, like soil type soil-fertility and rotations followed. Low-lying areas or other such features are also shown in the map. Then based on the previous crop history, land-capability classification is done and is also shown in the map.
2. *Inventory of farm resources*. Every asset on the farm, ranging from hand-tools to sources of power, etc. are inventoried. It does not provide us with the picture of the resources as owned by a farmer, but we can also work out their use-patterns and their condition, i.e. whether they will have to be replaced or whether they will be sufficient for the new plan or some augmentation will be needed.
3. *Examining the existing organisation*. having prepared an inventory of the existing resources and their availability, what we are interested in is their use-

pattern within the framework of the existing crop mix, whether the resources are understocked or overstocked. A careful analysis of the restrictions and weaknesses of the farm organisation is made. The weaknesses maybe many, such as:

— Lumpy units may not be fully utilised.

— Irrigation may not be adequate.

— The peak-season labour requirements may not be met.

— Short-term capital may be inadequate.

It is not simply the weaknesses that are emphasised but the good points of the existing organisation are also to be highlighted. The existing organisation may be improved or new good items maybe adopted.

4. *Laying down restrictions and planning*: Restrictions maybe with regard to bullock-power availability, or in respect of putting some area under cotton, or vegetables for home-consumption, though such a change in the cropping pattern may not be profitable. Labour-requirements, power requirements, capital needs and new equipment needed are worked out and a suitable crop mix is adopted.

In complete farm planning, the goals of the farmer are also important. He maybe desirous of investing in farming and, at the same time, he maybe willing to construct a new house. His goals and opinions and risks are all studied and weighed, and a compromising decision maybe taken. Thus in a complete farm-planning, due consideration is payed to resource use, restrictions, importance of relationships among different enterprises and to the goals and managerial skills of the operator.

Farm Credit

Analysis for Credit

A banker's plan is something more than the farm management specialist's plan. A banker is more concerned with the recovery of the loan advanced by him to the farmer. A farm-management specialist's plan is all right, but a banker, if he is to call it his plan, must examine it, as not only the additional returns from the alternative plan, but also whether the farmer would be able to repay the loan from the additional returns or not.

The credit forms the total capital requirement minus the farmer's owned funds. Most of the banks expect the farmer to contribute 25% of the total capital requirement as margin money so that he has some stake in the business.

Season-wise Analysis

Because the loan is to be paid at the end of every season, repayments include:

(a) The total amount in the case of self-liquidating loans which are for short periods and for purchasing seeds, fertilisers, hiring casual labour, etc.

(b) The instalments of the medium and long-term loans which are not self-liquidating and are for constructing buildings, raising irrigation structures, purchasing implements, machinery etc.

Farm-efficiency Measures

Efficiency is the ratio of the output to the input. This concept is important as it shows how much profitable the farming business is. Various measures to explain the efficiency of the business as a whole, and in parts, are available. When examined together they help to point out the weaknesses in the farm business and provide a guideline as to which part of the business deserves special attention for making improvements. In a particular situation due importance is given to a particular measure.

For instance different measures would be adopted for indicating the volume or size of the business, aggregate earnings from particular factors or the business as a whole and returns per unit of a particular factor input. Further, the efficiency of a farm can be judged from the costs or returns or both. No single efficiency measure is so complete as to give a true picture of the entire farm business. These efficiency measures help to reorganise the same farm for which they are calculated but they must be used with great caution, while comparing different farms.

This is especially important in the developing countries where farming is of diversified nature and individual farm business has wide variations in respect of soil type, resource restraints, capacities and capabilities of the farmers to undertake risks, their attitude towards innovations, etc. Also, these measures suffer from limitations for making comparisons because of their changing prices, costs, and conditions of the farm business. Adjustments with the price and coat indices are necessary before such comparisons give any valuable information.

Measures of the size of Farms

Farm income is directly related to the size of the business but in no way it indicated whether the farming business is operating efficiently or not. The following are some important measures of the size of a business.

(1) *Total area of farm and the number of livestock.* The acreage as a measure of the size of the farm may be either the total land, or the area under crops. The acreage at its face-value carries no meaning. It is necessary, therefore that land be homogenised and standardised for variations in fertility status, irrigation and lack of irrigation, etc. by such standard measures as rental value, land revenue, etc. As in a given type of farming area, there are important crops common to all farmers; therefore, the area (standardised) under principal crop enterprises will help to introduce further refinements in this measure.

In the case of livestock farms, the number of animals can be used as a measure of the size of the business. Again, the breed differences between the herds of two farms make a great difference in the returns received by them. On farms having both crop and livestock enterprises, any of the two will not give the relevant size of the business. Both may be used to represent the situation, but it is not comparable.

(2) *Total capital managed.* The total capital managed is obtained by adding all the capital values of inventories, including land. Either of the closing and beginning inventories can be used. This measure is useful when all the farms are producing the same product and this has no meaning for different specialisations. One is also not sure at the face value of the capital investment whether it is nearer to the optimum required, one farm can be compared with the other for the size of business by the capital investment only if the two farms have the near-optimum investments. The difficulty here

is that many farms are understocked or overstocked and the comparison between them is meaningless.

(3) *Total inputs.* The level of inputs in a broad way determines the level of output (income(. The total inputs mean the operational expenses plus the fixed expenses- the rental value of land, the value of labour (of family and operator) and the interest on the capital. This is the most reasonable method. It can be used in similar situations to compare with the assumption that rates of rent, interest and wages would be the same for each comparable situation.

(4) *Gross income.* This relates to the farm output during the year. This is obtained by adding the home-consumed farm products (in Rs) and the increase or decrease in inventory to the gross receipts of the year. This is a good measure but its use is limited to the same type of farming.

(5) *Productive manpower units.* They refer to the total manpower input required for normal efficiency. This method is more appropriate where labour is scarce and is being used optimally. In a labour-surplus situation, this method has only limited value. Also, the measure is limited to the comparing of the similar type of farming situations. Again, the difference arises from the difficulty to see whether the farmers under comparison use the scarce labour where it adds to the maximum or not.

These measures are rough indicators of the size of the business and can be used to compare only similar situations at a point of time and for the same type of farming.

Cost Concepts

There are many cost concepts which are in frequent use in the farm management literature, i.e. Cost A1, Cost A2, Cost B and Cost C; working or variable or operational cost and fixed costs; machine costs, labour costs, livestock costs, crop costs, etc. a brief discussion of these cost concepts is given below:

(i) *Cost A1*: It includes the following 16 items of costs:

1. Value of hired human labour (permanent and casual)
2. Value of owned bullock labour
3. Value of hired bullock labour
4. Value of owned machinery
5. Hired machinery charges
6. Value of fertilisers
7. Value of manure (produced on the farm, and purchased)
8. Value of seed (both farm produced and purchased)
9. Value of insecticides and fungicides
10. Irrigation charges (owned and hired tubewells, pumping sets etc.)
11. Canal-water charges
12. Land revenue, cesses and other taxes
13. Depreciation on farm implements (both bullock-drawn and worked with human labour)
14. Depreciation on farm buildings, farm machinery and irrigation structures.
15. Interest on the working capital

16. Miscellaneous expenses(wages of artisans, cost of ropes and repairs to small farm implements)

(ii) *Cost A*: It is cost A1 plus

17. Rent paid for leased land

(iii) *Cost B*: It is cost A2 plus

18. Imputed rental value of owned land (less land revenue paid thereupon)

19. Imputed interest on owned fixed capital (excluding land)

(iv) Cost C: It is cost B plus

20. Imputed value of family labour

It can be seen that cost C is a very comprehensive concept. But the risk and uncertainty costs are not taken care of.

Variable Costs

These are the costs which are of the recurring type, and have to be incurred during every production period, e.g. seed fertilisers and insecticides. These are also known as operational costs or working costs.

— *Fixed costs*. These costs are of non-recurring nature including cost of the tractor and other machinery buildings, irrigation structures, livestock, etc.

— *Crops costs*. These costs refer to the value of seeds and plants, manures and fertilisers, insecticides and fungicides and irrigation charges.

— *Machinery costs*. These costs include the hiring charges of machinery, cost of fuel and lubricants, electricity bills, minor machinery repairs, depreciation and interest on machines and equipment.

— *Labour costs.* They include the wages paid in cash or kind to the hired labour and the imputed value of family labour used on the farm.

— *Livestock costs.* They include the costs of veterinary medicines and services, fodders, feeds, interest on the values of livestock and depreciation.

— *Land costs.* They refer to the rent of land owned and taken on lease.

— *Building costs.* They include the interest on the value of the building structure and depreciation on them.

3

Irrigation Management in Crop Production

Water is one of the most important inputs essential for the production of crops. Plants need it continuously during their life and in huge quantities. It profoundly influences photosynthesis, respiration, absorption, translocation and utilisation of mineral nutrients, and cell division besides some other processes. Both its shortage and excess affects the growth and development of a plant directly and, consequently, its yield and quality. Rainfall is the cheapest source of natural water-supply for crop plants. In India, however, rainfall is notoriously capricious, causing floods and droughts alternately. Its frequency distribution and amount are not in accordance with the needs of the crops. Artificial water-supply through irrigation on one occasion, and removal of excess water through drainage on another occasion, therefore, become imperative, if the crops are to be raised successfully. Water management in India, thus, comprises irrigation or drainage or both, depending considerably on the environmental conditions, soil, crops, and climate. It is a situation-oriented entity.

Water effects the performance of crops not only directly but also indirectly by influencing the availability of other nutrients, the timing of cultural operations, etc. Water and other production inputs interact with one another. In proper combinations, the crop yields can be boosted manifold under irrigated agriculture.

Water is a costly input when canals supply it. The constructing of dams and reservoirs, the conveying of water from storage points to the fields, the operating and the maintaining of canal systems involve huge expense. The misuse of water leads to the problems of water-logging, salt-imbalance, etc., thus rendering agricultural lands unproductive. Hence a proper appreciation of the relationship among soils, crops, climate and water resources for maximum crop production.

Taking the total geographical area of the country at 328 million hectares and the average annual rainfall at about 112 cm, the total annual precipitation in the country is estimated at about 3,700,000 million cubic meters. The south-westerly monsoon contributes over 80 per cent to the total precipitation in the country.

The Central Water and Power Commission, New Delhi, has estimated that of the total annual precipitation amounting to 800,000 million cubic meters seeps into the ground, about 1,700,000 million cubic meters flows into the rivers and the remaining amount of about 1,200,000 million cubic meters evaporates back into the atmosphere.

Sources of Water for Irrigation

Irrigation is the replacement or supplementation of rainfall with water from another source in order to grow crops or plants. In contrast, agriculture that relies only on direct rainfall is sometimes referred to as dryland farming

The water source for irrigation may be a nearby or distant body of liquid or frozen water such as a river, spring, lake, aquifer, well, or snowpack. Depending on the distance of the source and the seasonality of rainfall, the water may be channelled directly to the agricultural fields or stored in reservoirs or cisterns for later use. In addition, the "harvesting" of local rain that falls on the roofs of buildings or on nearby unfarmed hills and its use to supplement the rain that falls directly on farmed fields also involves irrigation.

Various types of irrigation techniques differ in how the water obtained from the source is distributed within the field. In general, the goal is to supply the entire field uniformly with water, so that each plant has the amount of water it needs, neither too much nor too little.

The water, flowing on the surface and that seeping into the ground, forms the two major sources of water for irrigating crops.

Surface-Water Resources

A large number of rivers of various potentials and discharges are spread all over the country. The rivers in the north, which originate from the Himalayas, are snow-fed and, thus, have less seasonal fluctuations in their flow than the rivers in the other parts of the country. The flow in the rivers of the central and southern parts of the country depends entirely on the monsoon. The rivers flow to the full during the rainy season (July to September) and their flow dwindles with the approach of the summer.

The surface-flowing water needs to be trapped in ponds, tanks, lakes or artificial reservoirs when it is available in abundance so that it can be fruitfully used for irrigation during the rainless period to meet the water needs of crops. Of the annual surface flow of 1,700,000 million cubic meters, only about 666,000 million cubic meters can be utilised for the purpose of irrigation owing to the physiographical limitations.

Ground-Water Sources

Substantial supplies are also available from ground-water sources. Of the 800,000 million cubic meters of rain-water that seeps into the ground annually, about 430,000 million cubic meters of it is absorbed by the surface layers of the earth's crust and, thus, can be utilised directly by the vegetation in the process of evapo-transpiration and growth. The remaining 370,000 million cubic meters of rain-water percolates deep into the porous strata of the earth's crust, representing the gross annual enrichment of the underground water. This ground-water is tapped by digging or drilling shallow or deep wells and is lifted by using mechanical devices for irrigating the crops.

A precise quantitative inventory regarding the ground-water reserves is not available. Organisation such as the Geographical Survey of India, the Central Ground-Water Board and the State Tube-Wells and the Ground-Water Boards are engaged in this task. It has been estimated by the Central Ground-Water Board that the total ground water reserves are on the order of 55,000,000 million cubic meters out of which 425,740 million cubic meters have been assessed as the annual recharge from rain and canal seepage.

The Task Force on Ground-Water Reserves of the Planning Commission has also endorsed these estimates. All recharge to the ground-water is not available for withdrawal, since part of it is lost as sub-surface flow. After accounting from these losses, the gross available ground-water recharge is about 269,960 million cubic meters per annum. A part of this recharge (2,460 million cubic meters) is in the saline regions of the country and is unsuitable for use in agriculture owing to its poor quality.

The net recharge available for ground-water development in India, therefore, is of the magnitude of about 267,500 million cubic meters per annum.

The Working Group of the Planning Commission Task Force Ground-Water Reserves estimated that the usable ground-water potential would be only 75 to 80 per cent of the net ground-water recharge available and recommended a figure of 203,600 million cubic meters per annum as the long-term potential for ground-water development in India.

Irrigation Development

Irrigation has been practised in India since time immemorial. Wells and tanks, as known today's as 'minor irrigation works', were constructed by several kings of ancient times for public welfare. Significant development irrigation, however, can be stated to have commenced from 1850 onwards, with large-scale or major irrigation projects. These activities have been further accelerated after Independence in 1947.

Today, India has about 34 million hectares of land under irrigation and this amounts to about one-fifth of the total cultivated area. Source-wise, crop-wise irrigated areas in the country are shown in Tables 1 and 2. Major irrigation projects, irrigating more than 0.10 million hectares, have been listed in Table 3.

With utilisation water resources, both from the surface flows and from the ground-water recharge, the Second Irrigation Commission has calculated that the ultimate area that can be irrigated is of the order of about 82 million hectares in the country. Till the end of the Fourth Plan, about 45 per cent of the surface utilisable flow and about 20 per cent of the usable ground-water was proposed to be utilised to irrigate about 34 million hectares.

Table 1. Net area irrigated from different sources in 1970-71

(Thousand ha)

State/UT	Govt canals	Pvt canals	Tanks	Wells	Other Sources	Total
1. An.Pradesh	1,579	-	1,112	509	113	3,313
2. Assam	71	291	-	-	210	572
3. Bihar	812	2	169	551	626	2,160
4. Gujarat	204	3	30	945	10	1,209
5. Haryana	951	-	1	575	5	1,532
6. Him.Pradesh	-	-	-	1	90	91
7. J&K	62	210	-	1	6	279
8. Karnataka	419	2	365	259	92	1,137
9. Kerala	201	10	13	5	142	431
10. Madhya Pradesh	709		130	562	78	1,480
11. Maharashtra	280	32	225	821	69	1,427
12. Manipur	-	-	-	-	65	65
13. Meghalaya	-	-	-	-	37	37
14. Nagaland	-	-	-	-	12	12
15. Orissa	231	32	583	45	258	1,1479
16. Punjab	1,286	6	-	1,591	5	2,888
17. Rajasthan	756	-	270	1,083	23	2,132
18. Tamil Nadu	883	1	898	774	36	2,592
19. Tripura	-	-	-	-	22	22
20. Uttar Pradesh	2,497	1	371	17	209	1,489
21. West Bengal	629	331	303	17	209	1,489
22. Andaman & Nicobar	-	-	-	-	-	-
23. Arunchal Pradesh	-	-	-	-	-	-
24. Dadra &Nagar Haveli	-	-	-	-	-	-
25. Delhi	14	-	1	34	-	49
26. Goa, Daman&Diu	-	-	-	-	8	8
27. Laccadives	-	-	-	-	-	-
28. Mizoram	-	-	1	-	1	2
29. Pondicherry	11	-	6	10	-	27
India (Total)	11,594	923	4,537	11,864	2,404	31,292

Table 2. Crop-wise gross irrigated area in 1970-71

(Thousand ha)

Crop	Total Area	Irrigated Area	% irrigated Area	Relative area in %of total irrigated area
Rice	37,387	14,917	49.9	38.7
Sorghum	16,994	626	3.7	1.6
Pearl-millet or bajra	13,192	514	3.9	1.3
Maize	5,836	925	15.8	2.4
Crowfoot millet	2,525	334	1.3	0.9
Wheat	18,139	9,829	54.2	25.5
Barley	2,558	1,328	51.9	3.4
Gram	7,847	1,224	15.6	3.2
Pigeon-pea	2,652	12	0.5	0.03
Sugarcane	2,650	1,929	72.8	5.0
Spices and condiments	1,873	660	35.2	1.7
Fruits and vegetables	3,598	956	26.6	2.5
Groundnut	7,480	378	5.1	1.0
Sesamum	1,874	44	2.3	0.1
Rape and mustard	1,400	321	22.9	0.8
Lineseed	1,203	17	14.1	0.04
Cotton	7,886	1,287	16.3	3.3
Fodder crops	7,385	1,347	18.2	3.5
Tobacco	448	101	22.5	0.3
Other crops	37,534	4,498	12.0	12.0
Grand Total (gross area sown)	167,412	38,552	23.0	100
Net Area Sown	141,161	31,292	22.2	81.1
Area Sown more than Once	26,251	7,260	0.8	18.9

Table 3. Important Irrigation projects in India

State	Name of Project	Irrigated Area (Million ha)
	During 1850 to 1950	
Andhra Pradesh	Godavari Delta System	0.35
Andhra Pradesh	Krishna Delta System	0.60
Bihar	Sone Canal	0.24
Punjab	Western Jamuna	0.52
Punjab	Sirhind Canal (Sutlej)	0.95
Punjab	Upper Bari Duab Canal (Beas)	0.40
Rajasthan	Ganga Canal (Sutlej)	0.22
Uttar Pradesh	Upper Ganga Canal	0.80
Uttar Pradesh	Lower Ganga Canal	0.60
Uttar Pradesh	Sarda Canal	0.80
Uttar Pradesh	Agra Canal (Jamuna)	0.18
	After 1950	
Andhra Pradesh	Krishna Delta System	0.60
Andhra Pradesh	Nagarjunasagar (Krishna)	0.81
Andhra Pradesh	Pochampad (Godavari)	0.23
Bihar	Gandak	1.31
Bihar	Sone Barrage	0.12
Gujarat	Kakrapara (Tapti)	0.23
Gujarat	Mahi Right Bank Canal	0.19
Gujarat	Narmada	0.40
Gujarat	Ukai (Tapti)	1.60
Madhya Pradesh	Gandhisagar (Chambal)	0.45
Bihar	Sone Barrage	0.12
Madhya Pradesh	Kotah Barrage (Chambal)	0.57
Madhya Pradesh	Rana Pratap Sagar (Chambal)	0.40
Madhya Pradesh	Tawa	0.32
Maharashtra	Bhima	0.12
Maharashtra	Jayakwadi (Godavari)	0.14
Kerala	Kallada	0.11
Karnataka	Ghataprabha	0.20

Karnataka	Malaprabha	0.12
Karnataka	Tungabhadra	0.41
Orissa	Hirakud (Mahanadi)	0.24
Orissa	Mahanadi Delta	0.65
Punjab	Beas Dam	3.24
Punjab	Bhakra Nangal (Sutlej)	1.23
Rajasthan	Rajasthan Canal (Sutlej)	1.06
Uttar Pradesh	Matatila (Betwa)	0.17
Uttar Pradesh	Ramaganga	0.69
West Bengal	Kanasbati Reservoir	0.38
West Bengal	Mayurashi	0.25
West Bengal	Damodar Valley	0.42
West Bengal	Durgapur Barrage (Damodar)	0.42

The programs for modifying weather, the desalinisation of sea-water and the National Water Grid, if enforced, will further increase the potential for irrigation manifold. Surface water for irrigation is obtained from flowing rivers, and from tanks, ponds, lakes or artificial reservoirs. The flows of rivers are directly diverted into canals, or high dams are built across the river to form first large canals for irrigation. The future development of irrigation aims at impounding the surplus flows of rivers by constructing dams for use during the dry periods.

Water from all these sources is conveyed to the field through lined or unlined canals, distributaries and minors through the final structure called outlet. All this conveyance system up to the outlet is built by the Irrigation Departments. From the outlet, water flows into small water-courses which are constructed, owned and managed by a group of farmers.

During its conveyance, there are considerable losses through seepage, percolation and evaporation. From a typical water-distributary system, the losses in the main canal vary from 10 to 15 per cent, and in the water-courses from 15 to 30 per cent. Thus the total losses from the source till the water reaches the farmers' fields may amount to 40-60 per cent.

Ground-water is tapped by digging shallow and large diameter percolation wells or drilling deep tube-wells and lifting it to the surface. Shallow wells derive their water-supply from the surrounding area through seepage, percolation, high-water table, etc. the deep wells depend for their water on aquifers, which may have their source at some distance.

River valleys, canal-irrigated areas, low-lying places, natural vegetation and trees growing luxuriantly are indications of the presence of ground-water resources. The rate at which the water can be pumped out from a well depends upon recharging rate which, in turn, depends upon the permeability of the surrounding area in the case of shallow wells and on the thickness and the magnitude of aquifers.

Utilisation of water resources. The scientific utilisation of water resources for crop production involves the consideration of the suitability of land and water for irrigation and then planning of crops and water-management practices commensurate with them. Water-management practices include irrigation and drainage. Irrigation comprises three fundamentals: how much to drain, how best to drain and how rapidly to drain under a given situation of soil, water, and crops. These problems are discussed in this chapter.

Soil-water Systems

Soil is a heterogeneous mass and consists of three phases, viz. The solid phase, the liquid phase and the gaseous phase. Mineral matter, consisting of sand, silt and clay and organic matter, forms the solid phase which serves as a framework (matrix) with numerous pores of irregular shapes and different sizes holding air and water in various proportions. Soil is a porous medium, and serves as a water reservoir or bank. Water is deposited in this bank as rain or irrigation, and plants withdraw it during their growth.

Water is retained by a soil particle in the form of a thin film around it, and in the numerous small pores of the soil matrix with forces, such as surface tension capillarity, cohesion and adhesion. The salts present in soil water further add to these forces by way of osmotic pressure. Plants, therefore, need to exert at least an equal amount of force for extracting water from the soil mass for their growth.

Immediately after rain or irrigation, water infiltrates into the soil and continues to move in the soil mass to deeper layers because of the gravitational force. The downward movement of water practically ceases after a certain time (normally after 48 to 72 hours). The water retained in the soil under this situation is termed 'field capacity' which forms the upper limit of the available soil moisture for crop plant.

In other words, any further addition of water will not be retained by the soil, but will be lost through deep percolation beyond the roots of a crop, thus making it unavailable for the growth of its plants. After the wetting of the soil, as evaporation and transpiration continue, the soil water goes on diminishing till a point is reached when plants are unable to extract it. The moisture content at this stage is termed 'permanent wilting-point' and this sets the lower limit of the availability of soil water.

In other words, any moisture below this point will not support plant growth. The range of soil water between the field capacity and the permanent wilting-point is termed 'available soil water for crop growth'. The available soil water-holding capacity increases mainly with the fineness of texture and the content of organic matter.

Availability of soil water for crop growth. Three classical hypotheses have been put forth for the relative availability of soil water in the available range.

(i) Water availability and, consequently, the crop growth is equal and uniform over the entire range from the field capacity to the permanent wilting-point. This holds good generally for perennial species, such as orchard and tree crops whose dense root mass permeates the soil matrix thoroughly.

(ii) Water availability and crop growth proceed uniformly from the field capacity to a certain critical point beyond which crop growth decreases rapidly till the permanent wilting-point is reached. This view holds good for most of the seasonal field crops maturing up to the seed stage.

(iii) The availability of water and the rate of crop growth decrease gradually as the soil water content decreases from the field capacity to the permanent wilting-point. This holds good generally for most of the forage crops and those grown vegetatively.

The effects of the crop and climatic factors on the availability of soil water can also be significant. The responses of crops to soil water are ultimately reflected through the plant-water status, and the atmospheric aridity acts as the primary driving force for the absorption of moisture from the soil. A given crop loses water at different rayes under different conditions of evaporative demands of the climate.

Soil for Irrigation

For most crop plants, except rice, the ideal soil for irrigation is that which is deep, without any water-table, has high water-holding capacity, infiltration rate and permeability, and low salt content. The loams and clay loams are generally good soils for irrigation, since the run-off, the number of irrigations necessary and the investment for drainage are low as compared with those in the case of other soils. Any soil can be put under irrigated agriculture permanently but only with due care and caution which, on certain soils, may be beyond economic limits. The heavy soils often need surface as well as subsoil drainage. The light sandy soils involve a high application of costly inputs besides a considerable wastage of water.

In the case of rice, the soils with low percolation rates are ideal for economising on water. Land submergence has been found to be a beneficial practice for obtaining high rice yields. But during this submergence, huge quantities of water, as high as 60-70 per cent, are lost through deep percolation. In India, nearly half the water resources are diverted to rice alone. The proper selection of land for growing rice is, therefore, vital for utilising the water resources efficiently and economically.

Water Absorption by Plants

A plant extracts water from the soil with the hairs on the absorbing surface of their elongating roots, through osmosis and diffusion. Inside the plant, water moves along the concentration gradient across the root tissues to the xylem vessels and from there upwards to the stem and the leaves. On the surface of a leaf, there are numerous openings, stomata, through which water is lost in the form of vapour on account of the incident heat energy. The accumulated water vapour around the leaf tissues is further carried away by the wind into the atmosphere.

The process of loss of water from the plants through the stomata and plant surface is known as transpiration. Huge amounts of water, practically all that is absorbed, is lost in this way by the plants continuously throughout their life. The quantity of water utilised in metabolic activities by a plant is insignificantly small as compared with the amount lost through transpiration. The primary cause of transpiration is the solar energy available for the vaporisation of water from the leaf surface. Water is pulled out as a continuous film from the root hairs to the leaf surface because of the sink strength of the atmosphere.

Stomata are photosensitive cells and are open only during the day in most of the plant species. They close at night and hence there is practically no nocturnal transpiration. The stomata regulate transpiration to some extent by closing their apertures when the soil is dry and there is shortage of moisture in the atmosphere. The exchange of gases for photosynthesis also takes place through the stomata and, hence, the closure of stomata means no photosynthesis, i.e. no growth and accumulation of dry matter.

Plant nutrients also move primarily through water in the plants and are translocated to different plant parts along with the transpiration stream. The closure of the stomata, therefore, also reduces the nutrient uptake.

Quality of Irrigation Water

In irrigated agriculture, the quality of water used for irrigation should receive adequate attention. Irrigation water, regardless of its source, always contains some soluble salts in it. Apart from the total concentration of the dissolved salts, the concentration of some of the individual salts, and especially of those which are most harmful to crops, is important in determining the suitability of water for irrigation. The constituents usually determined by analysing irrigation water are the electrical conductivity for the total dissolved salts, soluble sodium percentage, sodium absorption ratio, the boron content, pH, cations, such as calcium, magnesium, sodium, potassium, and anions, such as carbonates, bicarbonates, sulphates, chlorides and nitrates.

Water from the rivers which flows over salt-affected areas or in the deltaic regions has a greater concentration of salts sometimes as high as 7,500 ppm or even more. The quality of tank or lake water depends mainly on the soil salinity in the watershed areas and the aridity of the place.

The quality of water from ground-water resources, i.e. from shallow or deep wells, is generally poor under the situation of :

(i) high aridity - arid and semi-arid regions, with less than 45 cm of rainfall;

(ii) high water-table and water-logged conditions ; and

(iii) the vicinity of sea-water, as in the coastal region.

Ground-water in other areas is generally good. In saline regions, where the water from shallow aquifers is poor, the water from deeper strata may be good. Such water may be tapped by digging deep-tube wells.

The continual use of poor-quality water for irrigation water can be improved by adding certain substances, such as sulphuric acid or sulpher to reduce the pH to the desired level. Gypsum, which is available commercially, can be applied to the soil or mixed with water where poor-quality water, containing a high concentration of sodium salts, is used for irrigation. The permeability of soils can be improved by incorporating organic matter into them to facilitate the leaching down of the salts beyond the root zone.

Table 4. Salinity and sodium-hazard classes of irrigation water

Salinity class	Electrical conductivity micromhos/cmm at 25°C
C_1 Low	Less than 250
C_2 Moderate	250 to 750
C_3 Medium to high	750 to 2,250
C_4 High	2,250
C_5 Very high	5,000 to 20,000
Sodium class	**SAR**
S_1 Low	Less than 10
S_2 Moderate	10 to 15
S_3 High	18 to 26
S_4 Very high	More than 26

Practices for crop production under conditions of soil or water salinity are given below:

- — *Choice of crops and varieties.* A crop or a variety which is more tolerant to salt should be selected for growing under the cropping pattern adopted. Crops have been broadly grouped as follows, according to their salt-tolerance limits.
- — *Tolerant species* (over 5,000 ppm salts): Barley, sugar-beet, date-palm, rape, kale, cotton, Rhodes grass, sesbania.
- — *Semi-tolerant species* (2,500 to 5,000 ppm salts): Rice, sorghum, maize, barley, sunflower, lucerne, berseem or guar, safflower, onion, spinach, lettuce, carrot, cluster-been, wheat, pearl-millet or bajra, grasses.

— *Sensitive species* (below 2,500 ppm salts): Peas, cabbage, vetch, beans, gram, peaches, grapes, orange, grapefruit, potato, tomato.

The limits of relative tolerance of crops to boron are as under :

— *Tolerant species* (2 to 4 ppm): Date-palm, sugar-cotton, lucerne, onion, turnip, cabbage, carrot and lettuce.

— *Semi-tolerant species* (1 to 2 ppm): Wheat, maize, barley, cotton, sunflower, potato, tomato, peas, beans and sweet-potato.

— *Sensitive species* (0.3 to 1.0 ppm): Apple, apricot, grape, orange, grapefruit, plum, pear, cherry and walnut.

Other cultural practices to ameliorate the poor quality of water:

— Emergence from the soil, the seedling phase and flowering are three very critical stages in the growth of most of the field crops in the presence of salinity. Frequent irrigation and in amounts in excess of the storage capacity of a soil are essential for maintaining low moisture stress and to leach down the salts.

— Row crops should be grown near the furrow bottom where salt concentration is low. The plant population should also be enhanced by increasing the seed-rate and by reducing the spacing to compensate for the poor performance of individual plants.

— Water, if only saline, can be used by diluting it with good quality water. If it is a solid soil, there will be a permeability problem. It is necessary to replace the sodium by calcium by applying suitable ammendments and organic matter for allowing more water to penetrate the soil, so that salts can be easily leached down beyond the root-zone. Adequate arrangements for drainage should be made to lower the water-table and to maintain a satisfactory salt balance in the root-zone. Fertiliser releasing nitrogen slowly with a calcium base may be preferred and applied in split doses by placement.

Planning Water Management

The water-management system on a farm includes farm inlet and outlet structures, conveyance lines, the laying out of irrigation and drainage ditches, the erecting of water-control structures, such as water gates, diversion boxes and drop structures, and the grading or the levelling of the land. It may also include the construction of storage tanks, the installation of pumps, etc.

Table 5. The proper assessment of availability of water, soils and crop factors

Water availability	Soil	Crop
Quantity per day	Texture	Type of root system
Rotation or turn period	Structure	Life-span
System and method of irrigation	Depth up to water-table	Consumptive water needs in relation to climate
Water quality class	Infiltration and permeability	Critical periods with respect to moisture
	Depth of a soil	Market value
	Slope of land	Yield response in relation to water-supply

After a proper assessment of these factors and topographic survey, a farm is divided into a convenient number of blocks and plots. Each block should have an independent irrigation and drainage channel and an approach road. There should be the minimum wastage of land under the field bunds and channels. The water flow in the channel should be non-silting and non-erosive.

Water can be conveyed from the storage point or the farm inlet to the point of use through earthen or lined water-courses or through metallic, canvas, cement or plastic pipes, hoses or tubes. Several lining materials of varying cost and durability are now available. These are mud, cement, bitumen, asphalt, bricks, plastic membranes, sodium compounds, etc. Conveyance losses are extremely high, and lining is essential for using water economically. The water-courses should be at a higher elevation than the field level.

Land Drainage

The provision of drains is a must on every irrigated farm for sustained farming. Damage caused by excess water and the consequent development of the salt problem is enormous and unnoticeable in early stages. Except rice, almost all crops receive a severe setback during their growth when excess water stagnates even for a period as short as 3 to 4 days.

In heavy-rainfall areas, there should be adequate arrangements for surface drainage for the speedy disposal of water. In high-water-table areas, there should be subsoil drainage. Spacing between the drains depends upon the permeability characteristics of the soil, excess water-tolerance capacity of the crop grown, the amount of water to be disposed of and its speed. The drains may be open or closed. The open drains are easy to construct and maintain, but they occupy considerable land. The closed drains are costly, but save land and, hence, are desirable where the land is scarce. The closed drains may be of mole type or tile type. A mole drain is an unlined channel in the subsoil made with a tractor-drawn mole plough. This type of drainage is suitable only on heavy clay soils. These drains last for two to three years. The installation of the mole system is cheaper than the tile system.

Tile drainage is rather an expensive system but it lasts several decades. The investment varies with the area to be drained, the depth of installation, the quantity of water to be handled and the type of soil, and the cost may range from Rs 1,500 to 3,000 per hectare. A tile is made of clay or concrete and it is about 30 to 50 cm in length and 7 to 12 cm in diameter. These are placed end to end, with a gap of two to three mm between them. Excess water enters the system through this space and is conveyed along the gradient.

The preparation of land for irrigation also includes grading or levelling. Several implements, such as scrapers, floats and land planes are now available for moving the earth and smoothening the fields. These can be operated with animal power or tractors.

Principles of Irrigation

The term water requirement includes the amount of water for meeting the needs of evaporation, transpiration and metabolic activities (all together known as consumptive use), losses during the application of irrigation water and water needed for special operations, such as land preparation, transplanting, the leaching of excess water, etc.

Water requirement WR = Consumptive use of water A +
Application losses B + Special needs + C

The entity A, viz. the consumptive use depends upon the factors of weather, the type of crop canopy, the soil moisture status and the stage of the crop. The entity B, i.e. the application losses, depends upon the type of the irrigation system, the soil texture and structure, and management practices. The entity C, viz., special needs, depends upon such factors as the soil moisture status of the soil as well as water, and the nature of the crop species grown.

Recent researches have revealed that meteorological factors play a very prominent role in deciding the magnitude of the consumptive use of water which forms a major part of the value of water requirements of crop plants. Evapo-transpiration involves the conversion of soil moisture into water vapour and requires about 540 calories per gramme of water. The primary source of energy is the sun. The role of wind is two-fold. It carries water vapour away from the leaves, steepening the gradient (sink strength) or it may bring additional energy known as 'advective energy'.

The advective energy can increase or decrease the consumptive use of water in a given locality. If there is a mass of active green vegetation, covering the ground completely and growing extensively under a non-limiting water-supply, the quantity of water evapo-transpiration by that mass of vegetation per day is independent of that species. Water loss by crop plants is, thus, primarily a physical phenomenon governed by the available solar and advective energy and the sink strength of the wind. The consumptive needs of water by a crop cannot be reduced and no moisture economy is possible in the case of this item on a field scale, if the aim is to obtain the maximum yields.

The entity B in the expression of water requirement comprises application losses during the conveyance of water through seepage and deep percolation. The irrigation efficiency depends upon the soil texture. The finer is the soil texture, the lower are the application losses and the greater is the irrigation efficiency than surface irrigation. The management practices also influence the irrigation efficiency. The levelling of the land properly, the selection of the optimum plot size and the stream size and the proper scheduling of irrigation depth increase the efficiency. The application losses can also be minimised by using water-control structures, such as water gates, diversion boxes and by lining the canals.

The entity C, i.e. special needs, depends upon the nature of the crop grown, the salinity status of the soil or water or of the both, the soil moisture status at the time of land preparation. More water is required for crop production if the transplanting of the seedlings is involved, if the soil or irrigation water is saline or if the soil is dry and hard before preparing the land for sowing.

The best, most rapid and most reliable method of finding out the consumptive use of water is with an open-pan evaporimeter which assesses all meteorological factors in an integrated way. The consumptive use of water is 0.7 to 8 times the value of the US class A open-pan evaporation when the soil moisture is adequate and the crop canopy is full and in an active stage of growth. A sunken screened open-pan evaporimeter furnishes directly the values of consumptive use of water by crop plants during their active growth.

The efficiency of field irrigation can be determined by measuring the quantity of irrigation water applied and stored in the root-zone. If it is not possible to measure the efficiency, the following broad values are suggested as a guide for surface-irrigation methods.

Soil class	Irrigation efficiency (%)
Sandy	60
Sandy loam	65
SandyLoam	70
Clay loam	75
Heavy clay	80

The quantity of water needed for special operations, such as land preparation, leaching down salts and transplanting, can be calculated by actual determinations in each case. The following values are indicated as a guideline.

25 to 40 per cent of the consumptive use 75 to 100

Operation	Details	Quantity of water (mm)
Land preparation	Light soil	50 to 60
	Heavy soil	70 to 80
Puddling and transplanting of rice	Light soil	300
	Heavy soil	200
Other crops	75 to 100	
Leaching down of salts	Low salinity	10 per cent of the consumptive use
	Medium salinity	20 per cent of the consumptive use
	Heavy salinity	25 to 40 per cent of the consumptive use 75 to 100

To find out irrigation requirements on a canal system, a period of 10 to 15 days is chosen. The values of irrigation needs during this period are worked out by subtracting the values of effective rainfall, ground-water and dew from the total water needs. These periodic values, when totalled for the entire period, furnish the total irrigation needs in a locality after making the correction for irrigation efficiency.

In the case of rice, the percolation need are worked out with the drum-culture technique and added to the values of consumptive use to arrive at the value of the total water needs. Subtracting the values of effective rainfall, ground-water contribution and dew from the total needs, the net irrigation needs would be obtained, when divided by the irrigation efficiency, the values of gross irrigation needs will be obtained.

In the case of the drum-culture technique, two drums, one with the bottom and the other without the bottom, are installed in the field. These drums are filled with water maintaining its level equal to that in the outside field. Daily, the difference between the water levels in the drums with and without the bottom gives the values of seepage losses. Differences between the two daily consecutive readings in the drum with the bottom furnish the values of the consumptive use of water. A third drum with its bottom, but with an overflow device is also installed for knowing the effective rainfall. The differences between the water-level in this drum and that in the drum without the bottom furnishes the values of ineffective rainfall.

Time of Irrigation

A crop should be irrigated before it receives a setback in its growth and development. There are several devices and methods, such as the use of blocks of Plaster of Paris, tensiometers, neutron moisture meter, soil-sampling, etc. for scheduling irrigation. But these devices and methods are either laborious and costly and are suitable only for experiments.

A careful observation of the system on leaves is not of practical use in the case of field crops. Plants growing on a slightly higher level or on a spot where the texture of the soil is relatively light in relation to the surrounding soil will show the symptoms of wilting earlier than the plants growing in the rest of the field and will furnish advance intimation for the application of irrigation. It is also possible to prepare a special spot of about one-to-two-metre cube by mixing a sand with the original soil. Plants growing on this soil-cum-sand mini-plot would indicate the irrigation need earlier than the rest of the field. This is the most inexpensive, simple and practical method suited for any crop and for any soil, except sandy soils.

Measurement of Irrigation Water

The application of precise quantities of irrigation water not only ensures a high efficiency of water use by the crops, but also reduces nutrient losses through leaching and results in better aeration of the soil. Irrigation water is measured under two conditions, viz. at rest and in motion. The units commonly used for expressing the volume of water at rest are litres, cubic metres, hectare metres, etc. The rate of flow is expressed in terms of units of volume per unit of time, e.g. litres per second, cubic metres per minute, hectare metres per day, etc.

The rate of flow of small streams on a farm can be directly measured by using a container or a bucket. The time needed to fill the container gives the rate of flow. This method can be used to determine the discharge rate of pumps and other water-lifting devices. The rate of flow of water into or out of a reservoir of known dimensions can also be directly made just by measuring the change in the water level of the reservoir with respect to time. Several types of water-meters are commercially available. They can be used for measuring the volume of water directly.

The rate of flow of water at a point in a straight ditch or in an open channel can also be determined by multiplying the cross-sectional area of water at that point by the average velocity. Under field conditions, it is impossible to maintain a uniform cross-section and, thus, a uniform velocity at various points.

The other devices commonly used for measuring irrigation water on a farm are orifices, weirs and flumes of various predetermined shapes and sizes. In most of these devices, the head of water on the upstream side is measured. The rate of flow is calibrated to it. Standard formulae, tables and calibration charts can be used to calculate the discharge of water through these devices. The commonest among these devices are the 90° V-notch weir and the Parshall flume.

Depth of irrigation

The quantity of water needed for net irrigation on different soil types per metre depth of soil profile at 50 per cent of soil-moisture availability is as follows :

Soil type	Irrigation depth (mm)
Sandy	25 to 50
Sandy loam	45 to 80
Sandy loam	70 to 110
Clay loam	80 to 120
Heavy clay	100 to 140

The depth of the soil to be wetted by irrigation depends upon the extent of the root-system of different crops. In the case of annual crops, the root-system is a dynamic one, growing to its full depth till about the flowering stage. In the case of perennials, the root-system is, more or less, stable in depth after about five years. The depth of irrigation, therefore, should be decided by taking into account the root-system of the crop. When rains are anticipated, it is better to under irrigate the field crop to accommodate the rainfall and reduce the amount of irrigation water to be applied. Important crops can be classified as under with respect to their maximum root-system on a normal well-drained soil.

Recent researches have indicated that a greater frequency of irrigation is more conductive to higher yields than a lower frequency. This means that it is advisable to apply light and frequent irrigations instead of heavy irrigations at longer intervals. The advantage from this practice is due to the increased uptake of nutrients, which are normally in the surface layers of the soil profile, and also due to smaller losses through leaching.

The new strains do not need higher quantities of water during their entire life than the old ones. But the new strains do require timely irrigation, since their productivity per day is very high as may be seen from Table 6.

Table 6. Productivity of old and new strains of crops

Crops	Varieties	Approximate duration (days)	Grain yield (kg/ha)	Productivity (kg/ha/day)
Rice	Old, tall	150	3,000	20
	New dwarf	150	4,500	45
Sorghum	Old	125	2,500	20
	New hybrid	110	4,500	40
Pearl-millet	Old	100	2,000	20
	New hybrid	95	4,000	42
Maize	Old	110	2,200	20
	New hybrid/ composite	110	5,000	45
Wheat	Old, tall	140	2,500	18
	New, dwarf	130	5,200	40

In other words, the new strains are more sensitive to moisture stress and delayed irrigation and moisture deficiencies which hit them more severely, resulting in heavy losses in yield. Table 7 shows what difference is made by four irrigations, timely and untimely, to a crop of wheat. The quantity of water applied (delta) in both the cases was nearly the same. Similarly, fertiliser applications do not increase the potential water needs of a crop for all practical purposes. But they increase the water-use efficiency by enhancing production. Nutrients in fertilisers need to be dissolved in water. They must move down from the soil surface to which they are generally applied and be distributed in the root-zone.

Table 7. Number of irrigations during crop age (days)

0-60	60-90	90-120	Yield (q/ha)	Relative yield
—	2	2	37.0	100
1	1	2	44.0	119
2	1	1	51.3	140
2	2	—	48.5	131
2	—	2	52.3	141

Soil-moisture Measurement

The knowledge of amount of moisture present in the soil is essential for deciding the time and the quantity of water to be applied. Several methods are known, and the choice depends on how precisely and quickly the information is needed, and on the availability of the requisite instruments.

Field Methods

Field methods yield immediate information, but the accuracy is just satisfactory for practical purposes. A farmer can assess the moisture status in the root-zone by inserting an iron probe or a tube. Resistance for forcing down the iron probe depends on the moisture content of the soil profile. In a wet soil, the probe will be forced down more rapidly and easily than in a dry soil. If a hollow tube is used for soil-sampling, the moisture can be estimated from the colour and the dry soil and will easily form a ball. By experience, these methods can be used fairly precisely for scheduling irrigations.

Laboratory Methods

Laboratory methods furnish precise values but involve costly devices and may be laborious and time-consuming. In the case of the standard gravimetric method, soil moisture is estimated by taking a sample with an augur and determining the loss in its weight when dried in an oven at 105°-110°C to a constant weight. The loss is expressed on the dry-weight basis. When there is no oven and electricity is not available, the soil sample can be dried by burning it with benzene, alcohol, or spirit and the loss in weight is recorded. This method, however, needs calibration with the gravimetric method.

The moisture percentage, expressed as percentage on the oven-dry basis, is converted into the total available moisture on volume basis by multiplying it by its bulk density. The bulk density is the dry weight of the soil per unit of its volume. Thus, if a soil has 20 per cent moisture on the oven-dry basis and weights 1.5 g per cubic centimetre, it contains 30 per cent water on volume basis or 30 cm per 100 cm depth of the soil when the area is the same.

At research stations, a record of soil-moisture changes can be obtained instantaneously by using devices, such as a tensiometer or a resistance block, which are rather costly. In research institutes, more efficient devices, such as the neutron moisture-meter or the gamma-ray method, can be used. These devices have a high precision but are very costly.

A tensiometer measures the force with which the surrounding soil is pulling the water held by it. A simple tensiometer consists of a porous porcelain cup, connected with a tube to a mercury manometer or a vacuum gauge. The tube is filled with water and the tensiometer is installed in the field with its cup at the desired soil depth in the active root-zone of the crop. When the soil moisture is depleted, water from the porous cup moves out, creating a vacuum in the system. The vacuum is recorded on the gauge or the manometer. With a multiple-cup tensiometer, the values in respect of the moisture status can be assessed with integration. A tensiometer works satisfactorily only in the wet range of available soil moisture (up to 0.8 tension).

A resistance block contains the electrodes embedded in a moisture-sensitive material, such as gypsum, nylon, and fibre glass. The block is installed at the desired depth in the soil and the wire leads are brought above the soil surface. When connected to a potentiometer, the block indicates the resistance to the flow of electricity, that varies with the moisture content of the matrix or the soil. In a wet soil, the resistance is low, whereas, in a dry soil, it is high. The device is calibrated against the gravimetric method for interpretation.

Irrigation Systems

The design, the equipment and the technique of replenishing the soil-water deficit by applying irrigation water is referred to as "irrigation system". The system adopted for irrigation must ensure a uniform distribution of water in the root-zone of a crop and high efficiency of water application i.e. the ratio of water stored in the root-zone to that delivered to the field, should be the maximum. There should be the minimum or no wastage of waste either through surface run-off or through deep percolation below the root-zone of a crop. The method used should be inexpensive and economically justifiable.

Several systems of irrigation are in vogue to suit different types of crops, topography, soil types, water resources, climatic conditions and costs. These systems are: surface-irrigation system, sprinkler irrigation system, subsoil-irrigation system, and drip-irrigation system.

Surface-irrigation System

In the surface-irrigation system, water is directly applied to the surface of the soil and is spread by gravity flow incidental to the slope of the land. There are several methods in this system, the commonest being flooding from a ditch, check basin, ring and basin, border strip and furrow. For irrigation with the surface system, fields are laid out every time before the crops are sown, since these layouts are destroyed during preparatory tillage. In order to achieve a higher water-application efficiency in the surface system of irrigation, the fields must be levelled as well.

The check-basin method of irrigation is widely practised in India, as it is well suited to all irrigable soils and to a variety of crop. Checks, rectangular or square, with sizes, varying from about 10 to 100 square meters or even more, are used. The basins are levelled in both directions. Slopes up to two to three per cent can be irrigated by using this method with a good control on irrigation water and high water-application efficiency. On steeper slopes, this method can be used after proper terracing. The limitations of this method are that it has too many ridges which not only occupy the land but also hinder interculture and harvesting with mechanical means.

The ring and basin method is commonly used for widely spaced orchard crop to irrigate individual trees or groups of trees. The important consideration in the case of the ring and basin method is to wet the soil only around the tree and not the entire land, thus economising on the water.

In the case of the border-strip irrigation method, the field is divided into long narrow strips, with small paralleled ridges on the sides. The strips are 3 to 10 meters wide, and across the width the strip is levelled perfectly. The length of a strip ranges from 50 to 300 meters or more, depending upon the slope, the rate of water-intake of the soil and the stream size. This method is also suited well to all irrigable soils and to closely spaced rowed crop and even to pasture crop. Slopes up to seven per cent can be irrigated when the pasture crop are grown. On steep slopes, this method can be used by proper terracing or trenching along the contours. For laying out border strips, the land needs to be graded uniformly to achieve a high water-application efficiency. Larger flows are required for irrigating border strips than those in the case of other layouts.

In furrow irrigation, water is applied to the field in furrow between ridges and the top of the ridge is not directly wetted. Crop, such as sugarcane, cotton, tobacco, potato, some vegetables and other widely spaced row crops, are irrigated by using the furrow method. The furrows can be along the slope when the level of the land is sloping gently up to about three per cent. When the slope exceeds three per cent and is up to about fifteen per cent, the furrows are laid out in graded contours.

Water distribution can be controlled well to achieve uniform application and the consequent high efficiency. The length of the furrows varies with the soil type, the slope and the quantity of water to be applied, and may vary from 10 meters to 1,000 meters for different situations. Under limited water resources, alternate furrow irrigation is adopted to save water.

Small furrows known as corrugations are sometimes prepared in the case of border-strip method to increase efficiency and the uniformity of overflooding from the ditch method. Corrugations can be made by using an implement called 'corrugator'. This implement is suited to close-growing crops grown on a medium type of soil with the rather uneven topography.

Sprinkler Irrigation System

Sprinkler irrigation is a system in the case of which water pressure is applied to the surface of any crops or soil in the form of a thin spray from above. This method is advantageous, as water can be applied at any controlled rate and uniform distribution and high efficiency can be ensured. This method can be adopted in the case of all crops and is very popular in the case of cash and some orchard crops.

The sprinkler system is specially suited to shallow sandy soils of uneven topography, where levelling is not practicable, and in areas where labour and water are scarce. On some soils with salinity problems, the sprinkler system is advocating for leeching of salt more effectively, emergence and to secure quicker and better growth of plants. The sprinkler system is also used for cooling the crops during high temperatures and for frost control during freezing temperatures. The pesticides, weedicides and fertilisers have also been applied successfully by adopting the sprinkler irrigation system.

A typical sprinkler system consists of a pump to lift and convey water under pressure, pipes or tubing for conveyance of water the sprinkler heads or nozzles, the risers which connect the sprinkler heads with the pipeline. Based on the equipment with which spraying is done, the sprinkler system has been classified as rotating-head type and perforated-pipe type.

The sprinkler system is also classified on the basis of the portability of the equipment as (i) portable, (ii) semi-portable, and stationary or permanent. The stationary type is more expensive than the portable one. In the portable and the semi-portable systems, again there are manual (hand-moved) systems or power-moved systems. These are self propelled sprinkler systems which move laterally or radially around central-pivot feeding-line. These portable systems can be designed to cover any area ranging from 3 to 4 hectares to 50 60 hectares or more. There are boom type sprinkler systems which employ one boom sprinkler head on each lateral. These systems irrigate an area of 75 to 100 meters radius, depending on the nozzle size and the pressure. The boom-type sprinklers are commonly used for tall crops and orchards.

In India, the portable systems, which are hand operated, are available commercially. They involve an investment 15-20 thousand rupees to install a system for irrigating five to ten hectares in one crop season.

Subsoil-irrigation System

In the subsoil-irrigation system, water is applied into a series of field ditches deep down to the impervious layer. It then moves laterally and then vertically through capillaries and saturates the root-zone. A continuous supply of moisture in the crop conditions created by the ponding of irrigation water on the impervious layer. In artificial sub-irrigation, which is also known as sub-surface irrigation perforated or porous pipes are laid underground in the vicinity of the root-zone and water under pressure is distributed through these pipes. This system is adaptable on soils having an impervious layer at a suitable depth in the sub-stratum. The initial investment and maintenance are prohibitive. The system is, however, very efficient, since the water losses through evaporation from the wet surface can be minimised. This system is practised at present in Gujarat and Kashmir for growing cash crop on the sandy-loam soils.

Drip-irrigation System

Drip-irrigation, also termed 'trickle irrigation', involves the slow application of water, drop by drop, as the name signifies, to the root-zone of a crop. The method was initiated in Israel and is now being tried in other countries. In this method, water is used very economically, since losses due to deep percolation and surface evaporation are reduced to the minimum. This method is, therefore, very much suited to arid regions and is being followed for irrigating orchard crops at present. The successful growing of orchards even on saline soils has been made possible by the drip system irrigation. The system can also be used for applying fertilisers in solution.

The equipment consists of a pumping-unit to create a pressure of about 2.5 kg per square cm, pipe-lines which may be of PVC tubing with drip type of nozzles or emitters, and a filter unit to remove the suspended impurities in the water. The amount of water dripping from the nozzles can be regulated, as desired, by varying the pressure at the nozzles, and the size of the orifice of the nozzles. Work is in progress in India to design, and adapt drip irrigation to conditions in this country. The initial high cost of the equipment and its maintenance are the major limitations of this system. It may, however, work out to be cheaper than the sprinkler system, especially for orchards and other widely spaced crops.

Water-lifting Devices

There are several devices for lifting water from low-level water reservoirs, such as well, lakes and rivers. Earthen pots, swinging buckets, wooden scoops, the Archemedian screw and picota are some of the ancient devices worked with manual labour and are well known in this country. They are suitable for low lifts. The discharge varies from 6,000 to 12,000 liters per hour. The mhote and the Persian-wheel have also been used in India for centuries. They are operated with animal power and can lift water from depths of 20 meters or so, with a discharge varying from 4,000 to 12,000 liters per hour. Horizontal and turbine pumps worked with petrol, diesel or electric motors, are the commonest present-day devices for raising water for irrigation. Wind velocities are not high enough throughout the year in India to raise water with wind-mills.

The power of a pump (hp), required to lift water, depends on the quantity of water, the lift involved, the efficiency of the pump and the frictional losses in the pipe-line. It is given by the relationship,

$$Hp = (62.4 \times Q \times H)/(550 \times E)$$

Where, hp is the brake horse-power of the pump;

62.4 is the weight in lb of the cubic foot of water;

q is the discharge in cubic feet per second;

H is the total head in feet;

550 is the foot-pounds per second; and

E is the efficiency of the pump.

The total head (H) is the difference in the elevation between the water levels and the area to be irrigated plus the fractional losses in the pipe-line. It represents the suction height plus the delivery height plus the resistance in the pipe due to friction expressed in feet. Where belt-drive or gears are used, the efficiency (E) is the product of the pump efficiency and the belt-drive or gear efficiency .

The installation of pumps should receive careful attention. The pump should be located as near the water source as possible. It should be free from floods and inundation hazards. The foundation for the base plate should be strong and rigid to absorb vibrations. The pump and the driving-shaft should be perfectly aligned and should be checked periodically.

The suction and delivery pipes should be supported adequately to avoid strain on the pipe casing. The piping should be adjusted with the minimum of bends. The pump should be protected from excessive pressure, and for the prevention of the back-flow of water by installing in the discharge line a non-return valve and the gate-regulating valve. A foot valve should be attached to the suction pump to facilitate priming.

Some of the important relations and equivalents in pump operations are as follows :

1 cubic foot of water = 62.4 lb

1 cubic foot per second = one cusec = 448.83 gallons per minute= 0.99 acre-inch per hour = 100 tons

1 acre-inch of water = 101 tons = 102.8 cubic metres = 22,660 gallons

1 cubic metre = 35.32 cft = 220 gallons = 1,000 litres

1 horse-power = 550 foot-pounds per second = 33,000 foot-pounds per minute = 746 watts

Irrigation practices comprise three interlinked components, viz. the quantity of water at each irrigation, the interval between two irrigations and the total number of irrigations in the life-span of the crop. The total number of irrigations received by a crop is governed by the availability of water, the needs of the crop, and the profitability of that crop.

Irrigation Practices for Major Crops

The interval between two irrigations depends primarily on the rate of soil-moisture depletion. Normally, a crop has to be irrigated before soil moisture is depleted below 50 per cent of its availability in the root-zone. The intervals are shorter during the summer than in winter. Similarly, the intervals are shorter in the case of sandy soils than in the case of heavy soil. When the water-supply is very limited, the intervals are prolonged. The crops are then irrigated only at critical stages.

The net quantity of water to be applied depends on the moisture deficit at the time of irrigation in the root-zone. The depth of irrigation applied is greater when irrigations are given at longer intervals. Similarly, the finer is the soil texture, the greater is the amount of irrigation applied on account of the higher water-holding capacity of the soil. The deficit (d) is estimated from the field capacity (FC), the actual soil-moisture content at a given time (Mc), the bulk density of the soil (Bd) and the depth of the root-zone (D) of the crops in question by the relationship :

$$d = FC - Mc/100 \times Bd \times D$$

Thus the actual irrigation practice for a particular crop will depend upon several factors from field to field and situation to situation. A good deal of discretion needs to be exercised for planning the irrigation schedule on a farm where there are several competing crops. Some guidelines are, therefore, indicated here, based on experiments conducted at several research stations in the country. These are applicable to tropical and sub-tropical conditions as obtaining in India.

Rice

Rice is grown both under lowland and upland conditions and throughout the year in some parts of the country. Under lowland conditions, the rice crop is generally transplanted in the puddled soil. Puddling disperses the soil and reduces percolation losses. It also helps to control weeds and create conditions of better nutrition. Direct seedling is also satisfactory, provided the weeds are taken care of either manually or by using chemical weedicides. For lowland rice, the practice of keeping the soil saturated or up to shallow submergence of about 5 cm throughout the growing period has been found to be the most beneficial practice for obtaining maximum yields. Shallow submergence is possible only if adequate and scrupulous care is taken in levelling the land. When water resources are limited, the land may be submerged at least during the critical stages of growth, viz. tillering and flowering, and maintained only saturated at other stages, thus economizing on water without seriously decreasing the yields.

For the crop grown during the rainy (kharif) season, when the weather is humid and evapo-transpiration rates are low and also when the water-table is near the surface, even maintaining the soil moisture near saturation is adequate. In other seasons, when weather is hot and arid, the practice of submerging the land is found to be more advantageous than when the soil is kept only saturated.

When continuous submergence is practised it is necessary to drain the soil two or three times during the period of growth, especially on poorly drained clayey soil. This practice is beneficial in removing toxic substance, such as the sulphides developed under reduced aeration. This practice should be synchronised with the period just following tillering and flowering. The drainage period should last from 4 to 8 days, depending upon the type of the soil.

The major portion of water applied to the rice crop, amounting to 50-75 per cent, is lost through deep percolation, which varies with the texture of the soil during the submergence of the land. The amount of water lost by the rice plants through evapo-transpiration is nearly equal to the evaporation from the U.S.W.B. Class A pan at a place.

Great economy in water use can be achieved in rice culture, if suitable measures are adopted to reduce the losses through percolation. The selection of heavy soils , the growing of rice in large and compact areas, instead of on small and scattered holdings, the providing of an impermeable layer, such as of cement, polythene, asphalt spray below the root-zone, helps to minimize deep percolation losses in the rice fields.

The amount of water required for growing rice in the country varies widely. The broad values under different conditions are as follows :

Total water need (mm)	Conditions
1,000 - 1,500	Heavy soils, high water-table, the presence of hard pan in the subsoil, short duration varieties, kharif season
1,500 - 2,000	Medium soils, kharif or early spring season

2,000 - 2,500	Light soil, long-duration varieties of rice grown during kharif or medium-duration varieties grown during summer, deep water-table

Under upland conditions, rice is grown by drilling the seed direct, and is matured on the rainfall during the season. The crop is irrigated whenever there is a break in the monsoon. The yields are lower than (nearly half) those under lowland conditions. The upland crop can be grown on medium to heavy soils under well-distributed rainfall of 1,600 mm.

Rice is an ideal crop for the areas likely to be flooded heavily by rains or by seasonally flowing river waters or by the inundation by lake shores. Special varieties, known as 'deep-water-rice' or 'floating-rice', are available. They grow very tall and very rapidly and can stand submergence up to three metres of water.

Wheat

Wheat is grown during the rabi season in the country, when the amount of rainfall is low and it is necessary to irrigate the crop for exploiting its yield potential. The optimum soil-moisture range for tall wheats is from the field capacity to 50 per cent of availability. The dwarf wheats need more wetness, and the optimum moisture range is from 100 to 60 per cent of availability. The active root-zone of the crop varies from 50 to 75 cm, depending upon the soil type. The critical stages during the growth are crown-root initiation (three weeks after sowing), flowering and grain development. Moisture stress at each of these stages can erode the yield by about 25 per cent.

Wheat is grown after the pre-sowing irrigation (rauni) when there is no adequate moisture at the time of sowing. Under Delhi and Punjab conditions, after sowing, the tall wheats need three irrigations, whereas the dwarf wheats can respond up to four to five irrigations. There is a likelihood of severe lodging in the case of tall wheats if they receive irrigation during the grain-development stage.

On sandy soils of Rajasthan and Gujarat, the number of irrigations varies from 10 to 15 on account of the higher evaporative demand of the climate. Water needs vary from 250 to 400 mm in northern India, and from 500 to 600 mm in central India. In Uttar Pradesh, a considerable area has a shallow water-table. One irrigation has been found to be adequate at the three-week stage for obtaining the maximum yields under such conditions. Corrugated border strips is the best lay-out for the efficient application of irrigation water to wheat.

Barley

Barley is similar to wheat in its growth habits, but can withstand more droughts because of its deeper and well-proliferated root-system. The active root-zone of barley extends to 60 to 75 cm on different soil types. The optimum soil moisture ranges from the field capacity to 40 per cent of availability. About two irrigations are adequate in the northern parts of the country on sandy-loam soils.

Maize

Maize is grown almost all over the country. The optimum soil-moisture range is from 100 to 60 per cent of availability in the maximum root-zone which extends from 40 to 60 cm on different types. The critical stages of growth are the early vegetative period (30 to 40 days after sowing) and tasselling (45 to 50 days). Moisture stress for a week during the early vegetative stage can reduce the yield by 25 per cent, whereas that during tasselling can affect it to the extent of 50 per cent.

The actual irrigation requirement of the crop varies with the amount of rainfall. In the northern parts, two or three irrigations (100 to 150 mm of water) are required to establish the crop before the onset of the monsoon. Afterwards, the irrigations are applied only if soil moisture falls below the optimum owing to long dry spells. At Arbhavi (Karnataka), three irrigations are found adequate during kharif, whereas five irrigations (330 mm of water) are necessary during rabi to obtain the maximum yield. At Bhubaneshwar (Orissa), 2, 11 and 18 irrigations are necessary to raise the crop during kharif, rabi and the hot-weather seasons respectively, the amounts being 100, 550, and 900 mm during these seasons. At Udaipur (Rajasthan), four irrigations (500 mm of water) are required for growing maixe during kharif.

Maize is very sensitive to excess water and, hence, it is advisable to make ridges in the field after its establishment. The submergence of the soil for three to four days during the vegetative or flowering period can reduce the yields up to 50 per cent.

Sorghum and other Millets

The millets are the hardiest among the cereals. Most of the millets are grown during kharif under rain-fed condition. They do respond, however, to irrigations profitably. The optimum moisture range is from the field capacity to 40 per cent of the availability.

The sorghum roots can extend down to 100 to 150 cm, but the other millets, such as pearl-millets, crow-foot millet and Italian millet, have shallow root systems, extending to 30 to 45 cm in the soil profile.

Pre-flowering and grain development are the two critical stages in respect of moisture. Wherever the facility exists, the crop should be irrigated at least at the critical stages to maintain optimum moisture in the root-zone during the dry spells.

Pulse or Grain Legumes

Grain legumes are the important sources of protein in Indian diet where the majority of the population is vegetarian. Important kharif legumes are cowpeas, green-gram (mung), black grain (mash), kidney-bean, and pigeon-pea. On account of their deep root-systems, the crops are generally grown under rain-fed conditions and rarely receive irrigation. When grown alone, one or two irrigation would be beneficial to boost their yields and income on account of their high market price and protein. The prominent grain legumes during winter are grain, lentil, pea, and Indian bean (Dolichos lablab). These crops are irrigated two or three times during their growth. Critical periods in their life-span of the legumes are early vegetative growth, flowering and pod development.

Grain legumes are extremely sensitive to excess moisture and hence should be grown on ridges wherever there is a likelihood of the soil becoming saturated for a prolonged period as a result of continuous and heavy rains. Pigeon-pea and gram are very hardy and can deplete soil moisture up to 1.5 to 2 meters deep and are, therefore, specially suited to water-scarcity areas.

Oilseeds

The principal oilseeds are groundnut, sesamum and niger during kharif and safflower, mustard, castor and linseed during rabi. The crops are generally grown under rain-fed conditions during kharif and often suffer a setback during their growth owing to ill-distributed rainfall. Maintaining moisture in the range of 100 to 50 per cent availability by one or two supplementary irrigation increases their yield more than double and pays rich dividends on account of their high price.

Groundnut is also grown during the winter in Andhra Pradesh and Tamil Nadu on red and laterite soils. Eight to ten irrigations of 50 mm each are applied at 10 to 15-day intervals during its growth period. Safflower, mustard and lineseed are grown alone or mixed with wheat or other cereals and receive three or four irrigations during their growth.

Branching (30 to 40 days from sowing) and flowering (60 to 70 days from sowing) are the critical stages with respect to moisture stress as well as with respect to moisture excess. It is always advisable to grow the with light ridging the field.

Safflower and mustard are relatively hardy crops and can stand more unfavourable conditions than others.

Cotton

Cotton is sown in India from April to October in different states, depending upon the species and varieties, soil types and the distribution of rainfall. It is a deep-rooted crop, extremely sensitive to excess moisture and lack of aeration. The optimum range of soil moisture for the crop is from the field capacity to 20 per cent of availability in 0 to 75 cm of the root-zone.

In northern India, the crop is sown in May after giving a pre-sowing irrigation (rauni) and then two or three irrigations are applied at intervals of 20 to 25 days during the pre-monsoon period. No irrigation is required during the monsoon. Water requirements under Delhi conditions vary from 400 to 500 mm. On sandy-loam soils of Ludhiana, two pre-flowering, two in-flowering and one post-flowering irrigations have been found to be optimum for obtaining the maximum yield. The number of irrigations is still higher in the arid regions of Rajasthan. In the black soils of Gujarat, two or three irrigations have been found to be adequate for growing cotton.

In Tamil Nadu, cotton is sown in September-October and thrives mostly on the north-east monsoon. Two or three irrigations are needed when the rainfall is not well distributed. At Arbhavi (Karnataka) about eight irrigations were required for a crop grown fro June to March on medium black soil. At Siruguppa (Karnataka) on black soils, two pre-flowering and four post-flowering irrigations have given the highest yields for a crop grown from August to March.

Optimum soils moisture is necessary during the stages of flowering and boll formation. The lack of adequate moisture at these stages results in the premature shedding of flowers and bolls, poor development of bolls, low ginning percentage and ultimately low yield.

Cotton is very sensitive to excess moisture and should be grown on ridges in humid regions. On heavier soils it is preferable to irrigate alternate furrows for maintaining better aeration. Jute. Jute needs continuously moist soils for the rapid elongation of the stem, the strands of which are used as fibre. The optimum moisture regime is from the field capacity to 70 per cent of availability in the maximum root-zone of the crops which extends to about 45 cm of soils depth. The crops can stand water-logging to some extent.

Sugarcane

Sugarcane occupies the land for about 10 to 18 months. The crops is grown under irrigation, except in some parts of Bengal, Assam, northern Bihar, and eastern Uttar Pradesh. The optimum soils moisture for sugarcane has been found to be 100 to 50 per cent range of availability in the maximum root-zone, extending up to 50 to 75 cm in depth. Since only vegetative growth is of economic importance, a good crops should never suffer from moisture stress during its growth. There is a linear relationship between the optimum soils moisture and the growth rate. Before the maturity of the crops, irrigation can be delayed for the accumulation of sucrose in the cane. The cane yield may go down as a result of this practice but not the sugar yield.

In the north, the crop is planted during February-March and is irrigated till the commencement of the monsoon. In Maharashtra, irrigation is required throughout the year. The levels of water requirements in different states have been broadly as follows: Bihar, 1,400 to 1,500 mm; Andhra Pradesh, 1,600 to 1,700 mm; Punjab, 1700 to 1,800 mm; Karnataka, 2,200 to 2,400mm; Madhya Pradesh, 2,000 to 2,300mm; Maharashtra, 2,800 to 3,000mm per year. The crop should be grown in furrows and later on, after four to five months, should be ridged. The ridge and furrow is the best layout for water economy.

Tobacco

Tobacco is grown for cigar, hookah, and bidi purposes. The optimum moisture regimes are from the field capacity to 70, 60, and 50 per cent of the availability, in these three cases, respectively. Cigar tobacco needs light and frequent irrigation during four months. For hookah tobacco, 12 to 13 irrigations of 50 mm of water each are required in Uttar Pradesh, Delhi and Punjab. In Gujarat, Maharashtra and Karnataka, the crop is grown on heavy soils under rain-fed conditions for bidi-making. When grown under irrigation, the crop should be raised on ridges. The quality of water is an important consideration in tobacco culture. Water, with a high chloride content, affects the leaf burning quality and hence, is unsuitable for irrigating tobacco.

Forage Crops

Fodders grow vegetatively and hence require frequent irrigations for maximum growth. The optimum moisture range is from field capacity to about 75 per cent of the availability.

Berseem is well suited to the regions in the north and require about 20 irrigations during its growth at intervals of about 20 days during December to January, at those of 15 days November, February and March, and at 10 days during September, October and April. The water requirement of the is about 800 mm during eight months of its growth. In the south, lucerne grows better and is perennial. It needs about 1,800 to 2,000 mm of water during the first year of growth. It is hardier than berseem. If the soil moisture is allowed to fall below 75 per cent of availability, the yield reduction in berseem, lucerne, cowpeas, sorghum and cluster-bean (guara) will be 30, 15, 15, 10 and 5 per cent, respectively. Hence, depending upon the availability of water, suitable forage can be selected for a given locality.

Vegetables

Most of the vegetable crops are short duration and grow quickly and need frequent irrigation for their maximum growth. The soil moisture should range between 70 and 80 per cent of availability in the maximum root-zone.

Potato is grown on sandy-loam soil in the north and need water at intervals of 10 to 12 days. It can best be applied in furrows between the ridges. In Haryana and Rajasthan, the irrigation intervals varies from eight to ten days, as are sandy loam. The crop needs about 500 mm of water during its growth. Stolonization and tuber-formation stages during 20-60 days growth are critical in their demand for water. Adequate water-supply should be ensured during these stages.

Onion and garlic are shallow-rooted crops and need very frequent irrigation as compared with any other field crop. Bulbing is the most critical stage from the point of view of applying water. About three weeks before maturity, the irrigation are delayed to enhance the keeping quality of the bulbs.

Tomato is grown in India throughout the year. The crop should be grown on ridges and irrigated at intervals of 10 to 12 days during the summer and of those of 15 to 20 days during the winter. Excessive irrigation, during ripening results in fruit-cracking and, hence, it should be avoided.

Cabbage, cauliflower and knol-knol should be grown on ridges. The optimum moisture regime is from 100 to 50 per cent of availability in these crops.

Radish, turnip, beet-root and other leafy vegetables need frequent irrigation to maintain the range of 100 to 75 per cent of availability in the top 30 cm of the

Water-melon and musk-melon are grown during the summer and need water at intervals of 8 to 10 days. These crops should be sown on the sides of ridges made 1.5 to 2 meters apart. Excessive irrigation should be avoided during the ripening stage, as it results in fruit-cracking. For the other of the cucubitaceae family, such as pumpkin, bottle-gourd, ridge-gourd, sponge-gourd, and cucumber irrigations should be applied at intervals of 10 to 12 days during the summer. The maximum root-zone of these lies between 40 and 60 cm in the soil profile.

Spices and Condiments

Important crops in this group are tumeric, ginger, chillies, ajwan, cumin and coriander. Tumeric and ginger should be grown on broad ridges and irrigated to maintain 100 to 70 per cent of available moisture in the maximum root-zone, i.e. the top 50 cm of the soil. Chillies should be grown on ridges and irrigated to maintain 100 to 50 per cent of available soil moisture in the active root-zone extending to about 60 cm in the . Coriander, cumin and ajwan are winter crops and need irrigation at intervals of 10 to 12 days on light soils and 15 to 20 days on heavy soils. The water requirements during their period of growth are around 500 to 600 mm.

Fruit-trees

Fruit-trees may be evergreen or deciduous. The deciduous trees shed their leaves during the winter and remain dormant for 3 to 4 months. The typical example is the grape-vine. The evergreen trees include citrus, mango, banana, chikoo, pineapple, papaya, date-palm, etc.

All trees require adequate soil moisture during their establishment period of 3 to 4 years. Soil moisture should be maintained in the range of 100 to 75 per cent of availability. Later on the full development of the root-zone down to 75 to 90 cm, the crops maybe irrigated when two-thirds of the available moisture is depleted during blossoming, fruit-setting and fruit enlargement. At other periods and during dormancy in the case of deciduous trees, irrigations may be applied when soil moisture almost reaches the wilting point.

Papaya and banana are shallow-rooted and short-lived species. They need to be irrigated when the available moisture in the top 30 cm of the soil layer is depleted to 80 per cent of availability. This will necessitate irrigation at intervals of 8 to 10 days during the rainless period in the tropical climate. The date-palm can withstand high temperatures and low humidity but needs regular irrigation during flowering and fruiting to produce good yields. In case of grapes too much moisture during the ripening causes the splitting and rotting of the berries. A slight stress during this period results in increased sucrose content and better colour.

Coffee

It is profitable to irrigate coffee after the cessation of the monsoon rains during flowering to avoid flower-shedding. Sprinkler irrigation has been successfully used to irrigate coffee during the post-monsoon period. This practice has doubled the crop yield.

Lawns and gardens

In lawns and gardens, the seasonal flowering plants need to irrigated very frequently to maintain the surface 30-cm layer moist. These plants should receive irrigation at intervals of 5 to 8 days, depending upon the severity of the climate. A lawn may be irrigated at intervals of 10 to 12 days in summer and at those of 15 to 20 days during the winter to maintain it fresh and green growth. In lawns, one has to resort to flood irrigation and, hence, proper consideration should be given at the time of planting to ensure that the site is perfectly level so that water will spread everywhere uniformly. The young shrubs and live fences should be irrigated along with the lawns. The fully developed trees may be able to tap water that has percolated deep. Hence, no special care is necessary for their growth though an occasional irrigation will be beneficial.

4

Management of Field Crop Production

Wheat Production

The Indo-Gangetic plains form the most important wheat area. The cool winters and the hot summers are very conducive to a good crop of wheat. Throughout India, only spring wheats are grown, although they are raised in winter. In the southern hills of Tamil Nadu, two crops of wheat are raised in a year. The first crop is taken between October-April and the second crop in May-September.

Climate

The annual rainfall in all the wheat growing zones ranges from 12. 5 to 100 cm. But most of this rain is received in summer or during the monsoon. In winter when the wheat is in the field, the rainfall ranges anywhere between 3 and 7 cm. To get good yield irrigation is essential and this enables the farmer to provide the necessary inputs. Irrigation facilities range very widely in different states. In Punjab, about 85 per cent of the wheat crop receives irrigation, whereas in states, like Maharashtra, Karantaka, Madhya Pradesh, only 13-25 per cent of the wheat area is irrigated. The high temperatures on both ends of the wheat season restrict the cultivation of this crop in India to the cooler months. High temperatures in September do not permit good tillering of the crop. They are also favourable to root rot and seedling blight fungi. The hot summer during the grain ripening period hastens the maturity of the crop.

Soils

Well drained loams and clayey loams are considered to be good for wheat. However, good crops of wheat are raised in sandy loams and the black soils also. In the past, it was considered that durum wheats were more suitable for cultivation in black soils than the aestivums. But this is not true with the present day aestivum wheats which are grown in all types of soils in the country.

Cultivation

The wheat crop requires a well-pulverised but compact seedbed for good and uniform germination. Three or four ploughings in the summer, repeated harrowing in the rainy season, followed by three or four cultivations and planking immediately before sowing produce a good, firm seedbed for the dry crops on alluvial soils. Timely cultivation and conservation of moisture are essential.

In the black cotton soil, blade harrow (bakhar) is used instead of the plough. One to three ploughings with an iron plough may sometimes precede the use of bakhar. For the irrigated crop, the land is given a pre-sowing irrigation (palewa or raund) and the number of ploughings is reduced. Where white-ants or other pests are a problem, Aldrin 5% or BHC 10% dust at the rate of 25 kg/ha should be applied to the soil after the last ploughing or before planking.

Sowing

Seed is sown by drilling or broadcasting. Sowing with seed drills, whether tractor drawn or bullock-drawn, will deposit the seed at uniform depth, give a uniform stand and lead to the early emergence of vigorous seedlings. In many places seed is sown by hand in furrows behind the plough, drawn by bullocks by the kera method. For irrigated timely sown wheat a row spacing of 15 to 22. 5 cm is followed, but 22. 5 cm between the rows is considered optimum spacing.

Under irrigated late sown conditions, a row spacing of 15 to 18 cm is the optimum. For dwarf wheats the planting depth should be between 5 and 5 cm. The planting beyond this depth results in poor stand. In the case of conventional tall varieties, the depth of sowing may be 8 or 9 cm. Under rainfed conditions after sowing, the furrows are left open, whereas in the case of irrigated wheat, the furrows are covered with a wooden plank (sohaga). Generally, deeper sowing is practised in rough, dry and light soils, whereas comparatively shallow sowing is done in moist soils. The transplanting of wheat seedlings can be done when sowings are delayed.

The seed used for sowing should have a good germination capacity and it should be healthy and free from seed borne diseases and seeds of weeds. It is advisable to sow certified seed, obtained from a good seed agency. For bold seeded and shy tillering varieties, e. g. ' Sonalika ', a seed rate of 100-125 kg per hectare is the optimum, whereas for the varieties with the medium sized grains, such as 'Kalyansona', 75-100 kg of seed per hectare. For late sown wheat, the seed rate is generally kept high at 125 kg per hectare to compensate for the lack of tillering. Under rain-fed conditions, a seed-rate of 100 kg/ha is recommended. The seed of loose smut-susceptible varieties should be given solar or hot water treatment. If the wheat seed is used only for sowing, and not for human consumption or for cattle feeding it can be treated with Vitavax.

In Rajasthan, Haryana, Delhi and western Uttar Pradesh, ear cockle and tundudiseases are common. In these areas, the galls should be removed by floating them in water and the seed freed from them should be sown after drying. The molyadisease caused by Heterodera avenae is an important nematode disease in Rajasthan and in parts of Haryana. Summer cultivation, rotation with non-cereal crops and the treatment of the soil with DBCP 60% EC at the rate of 30 l/ha in irrigation water minimises losses owing to this disease.

Irrigation

For wheat sown under irrigated conditions, 4-6 irrigations are required. The 1st irrigation should be given at the crown-root initiation stage about 20-25 days after sowing. Other irrigation should be given at late tillering, late jointing, flowering, milk and dough stages. Two or three extra irrigations are given where the soils are very light or sandy.

Under limited water supply, (a) if water is available for only one irrigation, it should be applied at the crown root initiation to tillering stage, (b) if water is available for two irrigations, the first irrigation should be applied between the crown root initiation and late tillering and the 2nd irrigation 7 or 8 weeks after the 1st irrigation (flowering stage), (c) if water is available for three irrigations, the 1st irrigation should be given at the crown root initiation stage and the 2nd and 3rd irrigations at intervals of 6 to 10 weeks. Thus the 3 irrigations should be given at crown root initiation, flowering and milk stages. Hoeing or interculturing a few days after the 1st and 2nd irrigations will break the crust and will also remove the weeds.

For controlling non-graminaceous broad-leaved weeds, a spray of 2, 4-d at 0. 4 kg a. e. /ha in 750 litres of water between 4 and 6 weeks after sowing is effective. Spraying before or after this specific period will be harmful to the wheat crop. If germinaceous weeds, e. g. Phalaris minor (foxtail weed) and wild oats are present, a pre-emergence application of tribunil at 1. 5 kg a. e. /ha in 750 litres of water controls them.

Harvesting and Yield

The crop is harvested when the grains harden and the straw becomes dry and brittle. The harvesting time varies from zone to zone and also depends whether the crop is grown under rain-fed or irrigated conditions. The rain-fed crop reaches the harvest stage much earlier than the irrigated crop.

In the peninsular zone, harvesting starts in the latter half of February and most of it is over by the end of this month or in the beginning of March. In the central zone, the crop is ripe by the end of February, and the peak season for harvesting is March. In the eastern zone, the harvesting starts in the latter part of March and is generally over by mid-April. In the north-western zone, the latter half of April is the peak harvesting time. In the hills the crop is harvested in may-June. The same wheat variety behaves differently in respect of maturity in the different zones. For example, 'Kalyansona' is harvested in peninsular India by the end of February, or the beginning of March, whereas the same variety planted at almost the same time in the northern hills is harvested in may-June. The high temperatures shorten the growing period. If the crop does not ripen before the onset of the hot westerly winds, the shrivelling of grains takes place.

Most of the harvesting in India is done with the sickle. Only in recent years, the harvesting is done in some parts of Punjab, western U. P., Haryana, etc., with combines on custom service basis. The grain is generally threshed by treading with cattle on a threshing-floor. The use of a stone roller, saw threshers and tractors expedite the operation. Winnowing is done generally with the winnowing baskets. Simple mechanical threshers, e. g. the Ludhiana thresher and the Sherpur thresher are used by some farmers. Threshers are very common in Punjab.

The yields under rain-fed conditions are generally low. They may be as low as 3 to 4 quintals per hectare in peninsular or central India. Under irrigated conditions, the yield as high as 100 to 115 quintals per hectare were reported whereas in the national demonstrations, yield up to 80 quintals per hectare was recorded. There was a false notion that best wheat yields were obtained only in north-western India. Recent experience with high yielding varieties indicate that similar yields can be obtained in eastern, central and peninsular zones provided the farmers grow the correct variety, and add to it all the needed inputs and follow the correct production technology. The all India average for wheat yields varies from 12 to 13. 8 quintals per hectare. The wheat straw is an important cattle feed in India. Generally the yield of straw from raw-fed or barani crops is about the same as the grain whereas from the irrigated crop it is one and a half to two times the grain yield. In the case of dwarf wheats, particularly triple dwarfs, the yield of straw is less than that from old varieties.

Rice

Rice (Oryza sativa) belongs to the genus Oryza. There are 18 valid wild species distributed mainly in Asia, Africa and America. Of the two cultivated species Oryza sativa and Oryza glaberrima, the former is cosmopolitan and the latter is confined in Africa. The extension of rice culture, extensive contacts between the people and selection pressure, both natural and artificial had led to the differentiation of races. Two major sub-species or races, i. e. indica and japonica, are recognised. All the rice varieties of India belong to the former group, though types resembling japonicas or with those of japonica type grains have also been reported from India and Nepal.

Rice cultivation in the world extends from 39 degrees S latitude (Australia) to 45 degrees N latitude (Japan) and 50 degrees N latitude (China) in India, it stretches from 8 degrees N latitude to 34 degrees N latitude. Rice is also grown even in areas below the sea-level, as in Kuttanad region of Kerala. It is also grown well at altitudes above 1, 979 m, as in parts of Jammu and Kashmir. It is cultivated as a purely rain-fed upland crop in West Bengal, Bihar, Uttar Pradesh, etc., where the monsoon is precarious and its distribution is often erratic. On the other hand, in parts of West Bengal and Bihar, shallow and deep water (4. 75-6 m) rices are also grown.

Rice shows considerable variability in grain size from 9 to 14 mm. It may be round, oval or elongated. The colour of the hull varies from straw to dark purple, brown or black, with various intergrades. Rice is normally white, but red and purple rices are grown in the country. The colour is due to either anthocyanin or non-anthocyanin. Variability is also considerable in other morphological characters, such as the root-system, plant height, tillering ability, leaf characteristics, panicle features and in physiological attributes, such as photosynthetic ability, degree of sensitivity to day length and temperature and the translocation potential.

There are two main seasons for growing rice in India, though three crops are taken in Tamil Nadu and Kerala.

The kharif season is characterised by a gradual fall in temperature, more numerous cloudy days, low light intensity, a gradual shortening of the photoperiod, high relative humidity and cyclonic weather. During rabi, there is a gradual rise in temperature, bright sunshine, near-absence of cloudy days, a gradual lengthening of photoperiod, and a lower relative humidity.

Rice cultivation is conditioned by temperature parameters at the different phases of growth. The critical mean temperature for flowering and fertilisation ranges from 16 degrees C to 20 degrees C, whereas during ripening, the range is 28 degrees C to 32 degrees C. Temperatures beyond 35 degrees C affect not only pollen-shedding, but also grain-filling. Broadly speaking, vigorous vegetative growth, moderately high temperatures are required, whereas for panicle initiation, slightly lower temperatures (20 degrees C to 22 degrees C) would be ideal. High temperatures and high light intensity adversely affect grain-filling.

Soil

The major soil groups where rice is grown are: (i) riverine alluvium, (ii) red-yellow, (iii) red loamy, (iv) hill and submontane, (v) terai, (vi) laterite, (vii) coastal alluvium, (viii) red sandy, (ix) mixed red and black, and (x) medium and shallow black.

The extreme soils conditions in which rice is grown are represented by acid peaty soils of Kerala (pH 3) and highly alkaline soils occurring in Punjab, Haryana and Uttar Pradesh (pH 10). For normal crop growth without any amendment, the pH range of 5 to 8. 5 is suitable.

Soil salinity affects the yield of many crops above 4 mmhos/cm conductivity, but rice, which is known to be a fairly tolerant crop, can grow normally in soils up to 6 mmhos/cm conductivity. The tolerance to the alkalinity is governed by the salinity of the soil, and of irrigation water, but generally exchangeable sodium percentage greater than 15 may adversely affect the crop growth.

Rice Culture Systems

The systems of rice cultivation in various rice growing areas of the country are largely dependent upon the rice growing conditions prevalent in the respective regions. The principal systems followed in India are 'dry', 'semi-dry' and the 'wet'.

The dry and semi-dry systems of rice cultivation are mainly confined to tracts which depend on rains and do not have supplementary irrigation facilities. The wet system is practised in areas assured of rainfall or by irrigation. In some areas, e. g. Assam, parts of Uttar Pradesh, West Bengal and Bihar, flooding is common during the rice growing season.

Dry and semi--dry systems

Basically, the steps involved are essentially the same under both these systems of rice-culture. The fields are ploughed and harrowed in summer for achieving the required-tilth. Farmyard manure is uniformly distributed 2-3 weeks before sowing. The seed is sown directly with the onset of the monsoon showers, either by broadcasting, dibbling behind the country plough or by drilling in lines. Line-sowing is preferable, as it ensures an adequate stand establishment and facilitates easy weeding and interculture. The reduced seed-rate requirement is another advantage. The row spacing may be suitably adjusted from 20 to 25 cm.

Under the semi-dry system, the rain-water is impounded when the crop is about 1-1/2 - 2 months old and thereafter it is converted into a wetland crop. By that time, major operations, such as weeding, interculturing and fertiliser applications might have been completed. 'Beushening' still prevalent in Orissa and Madhya Pradesh under this system helps to control weeds and adjust population. The latest thinking is to promote line-sowing, using a higher seed-rate so as to have a uniformly higher population density for effective competition from weeds and to use effective methods of inter culture to solve the weed problem.

Wet system

Under this system, the land is ploughed thoroughly and puddled with 3-5 cm of standing water in the field. The optimum depth of puddling is found to be around 10 cm in the clay and clay-loam types of soils. The primary objective is to obtain a soft seedbed for the seedlings to establish themselves faster, to minimise the leaching losses of nutrients and thereby increase the availability of plant nutrients by achieving reduced soil conditions which facilitates a better availability of nutrient elements, to incorporate the weeds and stubble into the soil and to minimise the weed problem. Puddling can be done with ploughs, tillers or tractors, depending upon their availability and soil conditions. The land is levelled after puddling to facilitate a uniform distribution of water and fertiliser.

Sowing

The seed-rate for direct sowing by broadcasting is 80 to 100 kg/ha and by dibbling it is 60-70 kg/ha. Only well-filled viable seeds should be used for sowing. Lighter seeds that float on a solution of common salt (1. 06 specific gravity) should be rejected. The choice of a suitable variety based on topo-sequence, soil type and duration is a prerequisite for obtaining high yield. Seed treatment with Agrosan at the rate of 100 g per 50 kg of seeds is generally recommended to prevent seed-borne diseases. The untreated seeds of high-yielding varieties should be soaked for 12 hours in a solution of wettable Ceresan (0. 1 per cent, i. e. 1 gm in one litre of water). The seeds should then be thoroughly dried in shade and used for sowing.

The nursery area required to provide seedlings for transplanting one hectare is roughly 1/10 of a hectare and the seed-rate is 40-50 kg/ha. To ensure rapid and uniform germination, selected seeds should be soaked for 24 hours in clean water, which should be drained away thereafter and the seeds be incubated in a warm, moist place for 36-48 hours to let them sprout for sowing in the nursery.

Interculture

In line-planted or drilled rice, weeding can be done with a hand-hoe or with rotary weeders. In a broadcast-sown crop, weeds pose a major problem and hand-weeding is still practised. The best time to weed the crop is three weeks after sowing or planting. Herbicides, such as Butachlor, 2, 4-D and Propanil, give a satisfactory weed control in rice and can be used as a tool for controlling weeds, when used at the recommended doses and at the right time. Irrespective of the method of weeding, it is ideal to maintain a weed-free condition up to 40-45 days after sowing.

Water Management

The water requirement of rice is higher than that of other cereal crops of similar duration. Figures ranging from 37 to 75 acre-inches of water have been reported from various locations in India, as the water requirement of rice, and this variation is primarily due to different soil and environmental conditions obtaining in different parts of the country. Losses due to percolation are more in submerged rice lands. In lighter soils, such losses amount to about 60 per cent of the total water requirement.

Losses due to transpiration account for about 40 per cent. Losses due to evaporation depend upon the climatic factors and range from 20 to 40 per cent. Considerable saving (30-40 per cent) in water is possible when shallow submergences (2-5 cm) is maintained from tillering to flowering and near saturation during other phases of growth.

Harvesting

Timely harvesting prevents losses in yield because of the shedding of grains. The later stage of grain-ripening is a dehydration process and maturity is hastened when water is withdrawn from the field at the hardening stage of the grains. The timely draining away of water makes the field hard to facilitate mechanical harvesting wherever this system is practised.

The harvesting of early and medium varieties, 25-30 days after flowering and 35-40 days in the case of late varieties is desirable. Field paddy should be harvested when the moisture content of the grain is about 20-25 per cent. This moisture content is to be reduced to 13 to 14 per cent before milling. The direct drying under the sun leads to an increased breakage of the grains during milling. Gradual drying in the shade is essential for better recovery in the mills. The produce after harvesting is known as rough rice, paddy or grain. The rough rice is milled before it is consumed. The husk or hull, the highly milled rice is white, translucent or opaque and is classified as head rice, brokens, screenings or brewers' rice, according to the size of the kernels.

Maize

The maize (*Zea mays L.*) plant is a native of the Americans. It was introduced into India by the Portuguese during the seventeenth century is often quoted, but the more recent evidences support the possibilities of its pre-Colombian introduction through the Atlantic Arabian trade route. The latter evidence is also supported by the more recent variability that has been observed in the maize collections made from the north-eastern Himalayan region. However, the historical records regarding the cultivation of maize in India date back only to the Maratha Empire. Maize in India is an important cereal, and both its area and production have steadily increased during the past decades. In terms of area and production, it ranks only next to rice, wheat, johar, and bajra. The states of Uttar Pradesh, Bihar, Madhya Pradesh, Rajasthan and Punjab account for over 75 per cent of the area and production of this cereal in the country.

Soil

Maize requires fertile, deep and well-drained soils. However, it can be grown on any type of a soils ranging from deep heavy clays to light-sandy ones. It is, however, necessary that the pH of the soil does not deviate from the range 7. 5 to 8. 5. Maize plants, particularly in the seedling stage, are highly susceptible to salinity and water-logging. Accordingly, the provision of proper drainage is essential for the successful cultivation of this crop. The light-sandy soils greatly facilitate drainage, but have a relatively poor water-holding capacity; on the contrary, very healthy soils, with excellent water-holding capacity, have relatively poor drainage. Hence, soils ideally suited for maize cultivation should have adequate drainage. Over 85 per cent of the maize acreage is sown under rain-fed conditions during the monsoon when over 80 per cent of the annual rainfall is received.

Climate

Maize is essentially a warm-weather crop. It is grown under extremely divergent climatic conditions in different parts of the world, ranging from tropical to temperate regions. It is widely cultivated from the sea-level up to altitudes of 2, 500 m. It can be successfully grown where the night temperature does not go below 15. 6°C (60°F). It cannot withstand frost at any stage of its growth. In India, its cultivation extends from the hot arid plains of Rajasthan and Gujarat to the wet hills of Assam and Bengal.

Most of the maize varieties grown in India are relatively early in maturity (80-95 days) ; hence to sustain the rapid rate of growth, an adequate supply of soil moisture is essential. It has been estimated that the maize crop requires about 50 per cent of it total water requirement in a short period of 30-35 days after tasselling. A lack of adequate moisture during the grain-filling stage adversely affects yield. Even though maize can be grown without additional irrigation in regions receiving about 60 cm of well-distributed rainfall, yet for obtaining the optimum yield, additional irrigations become necessary when the rains fail.

Cultivation

A good seedbed for maize should be fine but compact, and free from weeds. It is desirable that the previous crop refuse is buried under with a mouldboard plough. In due course, two or three ploughings with the wooden plough are given. In case the tractor is used, one ploughing followed by a couple of diskings is generally adequate.

For kharif cultivation, it is essential that adequate drainage is provided. It may be provided in the form of very shallow surface drains at 40-50 m apart (depending on the slope and the texture of the soil) across the slope and connected to a main outlet. Shallow drains do not obstruct the movement of cattle or tractors during cultivation. Surface drains should be provided before sowing.

For obtaining high yields, the maize crop should be heavily manured. Twenty-five to thirty cartloads of farmyard manure or compost should be ploughed into the soil before sowing. For hybrid and composite varieties of maize, 100-120 kg of nitrogen, along with 60 kg of P_2O5 and 40 kg of K_2O per hectare, is recommended. The precise level of application of phophorus and potash should be modified in the background of soil analysis. One -third of the nitrogen and total quantity of potash and phosphorus should be applied before sowing, while the remaining nitrogen should be applied as side-dressing at the Knee-high stage and at tasselling in two equal doses. In freshly levelled fields and soils with very light texture, a soil application of 10-20 kg/ha of zinc sulphate before sowing has also given good results. To control soil-infesting insects, 10-20 kg of 10% DDT or BHC per hectare may br mixed with the basic fertiliser application. Farmers planting local varieties may apply 40-60 kg of nitrogen to the soil per hectare.

There are three distinct seasons for the cultivation of maize : the main season is kharif; whereas its cultivation during rabi in Peninsular India and Bihar, and in spring in northern India is done. Higher yields have been recorded in the rabi and spring crops. The higher yields are primarily due to better waste management and a lower incidence of diseases and pests.

In most parts of India, maize during kharif is sown with the break of monsoon, the actual dates varying from region to region. It is sown in early March in north-eastern hills, in April to early May in nort-western hills, in May-June in Peninsular India, in the end of June to mid-July in the Indo-Gangetic Plains. The late sowing of maize may extend up to late August in certain irrigated tracts of Punjab. Spring maize is sown in late January to the end of February in Bihar and in the terai-belt of Uttar Pradesh. Rabi maize is generally sown in Bihar, Andhra Pradesh, Tamil Nadu and Karnataka in the ned of October to mid-November. Both the spring and the rabi crops are raised more or less, under irrigation.

Maize is sown in rows, 60-75 cm apart, whereas the plants in the row are spaced at 20 to 25 cm. A population of 60-75 thousand plants per hectare at harvest is required for obtaining the optimum yield. Sowing in rows is generally done with drill or by dropping the seed behind the plough. The practice of broadcasting, particularly under rain-fed conditions and for fodder maize is still prevalent in several parts of the country. Seventeen to 20 kg of seed for the grain crop and 35-40 kg per hectare for the fodder crop is needed.

In regions with about 60 cm of well-distributed rainfall during the growing season, any additional irrigation is not necessary. Inadequate soil moisture during flowering and post-flowering particularly during the grain-filling period will markedly reduce the yield. During the grain-filling period - the most susceptible stage, additional irrigation, if needed, should be applied. The total number of irrigations will depend on the rainfall-distribution pattern. The spring and rabi crops are entirely raised under irrigation; the number of irrigations may, however, vary from 5-10, depending upon the type of the soil and the prevailing temperature.

During the initial stage, the growth of the maize plants is suppressed by weeds. Weeding may be done between the rows with bullock- or tractor-drawn implements, whereas the weeding within the row is done by hand. Two to three weedings may be necessary, following which the crop is earthed up to provide for better standability. No intercultivation after flowering should be done, as it is likely to damage the lateral roots. Effective weed control can be obtained by spraying Simazine or Atrazine before the crop germinates. Before spraying, the fields should be free from weeds. For the fodder crop of maize, less of weeding is needed, since the soil surface is nearly covered by a dense population of the maize plants.

Harvesting

The maize crop sown for grain is harvested when the grains are nearly dry and do not contain more than 20 per cent moisture. The appearance of the plant may be misleading, particularly in the case of high-yielding hybrids and composites whose grains are dry, while the stalk and leaves may be still green. Ears are removed from the standing crop. Harvested ears are dried in the sun before shelling. In the case of the late-sown crop, farmers prefer to harvest the whole plants and pile them, and the ears are removed later. Maize stalks are used as cattle-feed and fuel. In fact, no part of the maize plant, even the cobs from which the grains have been removed, is left unused.

Maize grown for fodder should be harvested at the milk to early-dough stage; the earlier harvested crop is likely to yield less and have a lower protein content. For silage, however, the late dough is preferred. Both power- and hand-operated low-priced maize shellers are available indigenously. These shellers are considerably more efficient than hand-shelling or beating with sticks, the common practice in northern India.

Farmers using hybrid maize should not save their own seed for their next crop, as the advanced-generation hybrid seeds are likely to lead to a yield reduction of 25-30 per cent. However, farmers can save seed from composites and the open-pollinated varieties, when grown in isolation. At least seeds from 500 to 1, 000 ears of the best-yielding, normally-spaced plants resistant to prevalent diseases and pests should be bulked. Ears should be dried, shelled and treated with an insecticide and the treated seed is necessary, as the untreated seeds on ears are at times badly attacked by stored-grain pests and the germination is markedly reduced.

Considerable variation in grain yield is observed. The yield levels depend upon the variety, the amount of the fertilizer used, the rainfall pattern, etc.. Under irrigated conditions and recommended cultural practices, an average yield of 4 tonnes per hectare in the Indo-Gangetic Plains is not uncommon; in Peninsular India and at higher elevations, a mean yield of 5-7 tonnes per hectare has frequently been obtained. Under low-fertility and rain-fed conditions with poor-yielding varieties, a grain yield of about one to two tonnes/ha is obtained.

Barley

Barley (*Hordeum vulgare L.*) is an important rabi cereal of India, ranking next to wheat both in acreage and in the production of grain. The major portion of the grain produced is consumed by the poor classes either as flour (pure or mixed with that of wheat or gram) for chapati-making or as parched grains to make sattu. It is also used to prepare malt for manufacturing beer and whiskey and other products, such as industrial alcohol and vinegar. Malt syrup is also utilised for making candies, breakfast cereals, medicines and in textile industry. Malt sprouts and by-products of brewing are used as a feed for dairy cattle. The grains are also used for manufacturing pearl and powder products, which generally form the diet of sick people. Surplus grain provides feed for cattle and horses. Straw is also fed to cattle.

The total area under this crop is about 2. 9 million hectares, producing nearly 3. 1 million tonnes of grain. It is grown chiefly in the northern plains of the country, but its cultivation extends up to elevations of about 4, 575 metres (15, 000 feet) in the Himalayas. Its greatest concentration is in the states of Uttar Pradesh, Rajasthan and Bihar, which account for approximately 52, 18 and 11 per cent, respectively, of the total area. It is also grown to a considerable extent in northern Madhya Pradesh, Haryana, Punjab, West Bengal, Himachal Pradesh and Jammu and Kashmir. It is of minor importance in the south of 20°N, although it can be grown successfully wherever wheat can be grown.

Climate

Like wheat, barley does best in areas where winter is cool and the growing period lasts five months. It is ideally suited to areas where rainfall is either low or uncertain. In any given region, cultivated barleys mature 2 to 3 weeks earlier than wheat and thus escape serious drought, if encountered. Areas that are always warm and moist are not suitable for this crop. Frost at flowering and hailstorms when the grain development is complete damage the crop. Rains at the time harvesting cause dicolouration of the grains, making them unfit for malting. In some recent experiments, this crop has shown good adaptation to conditions obtaining in the coastal areas of West Bengal, which traditionally grow only one crop of paddy. It suitability in other non-traditional wheat and barley-growing states, such as Assam, Meghalaya, Arunachal Pradesh, Tripura, Manipur, the Andaman Islands and Orissa in the east; Andhra Pradesh and the hills of Kerala in the south is being determined.

Soils

This crop can be grown successfully on those soils which are suitable for wheat cultivation. However, it thrives well on well-drained and moderately rich loams. Severe lodging occurs when grown on extremely fertile soils. It is grown widely on a variety of soils, ranging in their texture from sandy to heavy loams in the Indo-Gangetic plains and on terraced slopes in the hills. Its cultivation also extends, although to a limited extent, to medium black soils of Rajasthan, Maharashtra and Karnataka. It is more tolerant to alkali and saline conditions than other rabi cereals and is grown extensively under such conditions in Rajasthan, Punjab, Haryana, Uttar Pradesh and Bihar. Recent experiment indicates that this crop can be grown successfully on coastal saline soils of Sunderbans in West Bengal and on saline-alkali black soils in the cannal-irrigated areas of northern Karnataka.

Cultivation

The cultivation practices for barley are the same as for wheat. Best yields are obtained from a well-prepared soil which is in good tilth and completely free from weeds. However, a very thorough cultivation is not needed. Three or four ploughings with a desi plough or one ploughing with an improved iron plough and one harrowing are generally regarded as adequate. For raising a crop under dryland farming, soil-and water-conservation measures, such as deep ploughing and desking after each rainfall, levelling and bunding should be done during the monsoon season to ensure proper soil moisture for crop growth. At the last ploughing, BHC (10%) at the rate of 20 kg per ha or Aldrin dust (5%) at 10-15 kg per ha should be applied to the soil to protect the crop against white-ants and other soil pests. The saline-alkali land should be prepared, planted and left undisturbed for about a week before sowing, so that the salts move from the lower soil layers to the upper soil-zone and the root-zone becomes relatively free from salts. Unlike wheat barley needs a slightly loose seedbed. On soils infested with nematodes, two to three deep ploughings during summer at ten-day intervals are recommended to reduce nematode population.

Sowing

The seeds are sown either by broadcasting or with a single-tube drill (pora, magha or sadde), operated in shallow furrows behind a plough. The sowing of seed in rows by using the Pora method or with a seed-drill ensures a uniform stand, less lodging and better weed control. The suggested spacing between rows is 22-23 cm under irrigation and 23-28 cm under the rainfed condition. The best depth for sowing is 3-5 cm under irrigation and 5-8 cm under rainfed condition, depending upon the initial soil moisture. In clay soils, or soils which have a tendency to crust, shaloow planting is preferable.

Manuring

The rainfed crop is seldom manured directly. It is the preceding crop in the rotation which receives the manure. Irrigated barley is often given a basal dose of eight to ten cartloads of farmyard manure or compost. Any subsequent application of a fertiliser, however, is not practised. Recent agronomic experiments indicate that barley responds strikingly to an application of 20-40 kg of N per ha under rainfed conditions, and to 40-60 kg of N per ha under irrigated conditions, depending upon the soil, climate and variety. The application of 20 kg of N per ha to the crop under dry-land farming led to the realisation of 2-6 q per ha of additional grain and an application of 40 kg of N per ha under irrigated conditions resulted in a yield of 7-9 q per ha more than that from unfertilised plots at different locations. Good response to phosphorus up to 20 kg of n per ha has been obtained under irrigated and timely-sown conditions, but only at some locations. Hence its application needs to be governed by a soil-test value.

To maximise the efficiency of nitrogenous fertilisers, the full dose of nitrogen should be placed 8-10 cm deep in the soil at sowing time under rainfed conditions. For irrigated barley, a split application of the half dose of nitrogen plus the full dose of P2O5 at sowing time and the remaining half dose of nitrogen with the first irrigation as a top-dressing is advantageous. Excess of nitrogen affects the malting and brewing quality of grain adversely. Superphosphate, however, helps to neutralise its deleterious effects.

Irrigation

Barley is grown either on conserved moisture or under restricted irrigation. It responds to irrigation in the drier areas and on soils with low moisture-retention capacities. Usually, it needs two or three irrigations. One or two extra irrigations will be required on sandy soils. If the supply of water is inadequate, its efficiency can be increased by a proper timing of its application at the critical stages of growth of the crop. If only one irrigation is available, its application near the tillering stage has proved very profitable. When two irrigations are available, one should be applied at the active tillering stage and the other at the flowering stage. On highly alkali-saline soils, frequent light irrigations give better results than a fewer heavy irrigations.

Harvesting, Threshing and Yield

Harvesting in Jammu, Punjab, Rajasthan and western Uttar Pradesh takes place from the third week of March to the middle of April. In Madhya Pradesh, southern and eastern Uttar Pradesh, Bihar and West Bengal, it starts a week to ten days earlier, and in Maharashtra, Gujarat and Karnataka by about the first week of February. In the hills, where it is grown during the rabi season, the harvesting time varies from the end of April to the end of May, depending upon the altitude. The spring-sown crop is harvested from the end of July to the end of September.

The method of harvesting is similar to that of wheat. The plants are uprooted or cut with a sickle when fully ripe. In order to prevent the loss of grain through shattering, it is advisable to cut the crop in the early hours of the morning. The thrashing in the plains is done by treading the dried produce under the feet of cattle. The grain is separated by winnowing.

The average grain yield of the 'dry' crop is about 700-1000 kg per ha, whereas that of the irrigated crop is about twice as much. Under favourable conditions of manuring and management practices, improved varieties are capable of giving five tonnes of grain per hectare under irrigated conditions and 2. 5 tonnes per hectare under rainfed conditions.

Oats

The oat (*Avena sativa L.*) crop is relatively a recent introduction and is of minor importance. It is grown in the rabi season mostly for fodder. Occasionally, in some areas it is grown for grain. It is also grown in pastures, cut for hay or green fodder and is also used for making silage. Its fodder and grain are highly nutritious and are especially preferred for milch cattle and draft animals, including horses. A very small portion of the oat grain is processed into food in the form of 'rolled oats' and 'oatmeal' for human consumption.

Climate and Soil

Like wheat and barley, this crop requires cool season for its proper growth. Its cultivation is largely confined to Uttar Pradesh, Punjab and Haryana. To a smaller extent, it is also grown in some parts of Maharashtra, Gujarat, Madhya Pradesh, Bihar, Orissa and West Bengal. The crop can be grown on all types of soils, the best results, however, are obtained on well-drained rich friable loams. The crop is grown primarily under irrigation. Only in moist and cool climate, it can be grown as a non-irrigated crop. It is grown in rotation with leguminous or non-leguminous fodder and grain crops. The crops in the proceeding kharif are maize, pearl-millet, sorghum, cowpea, cluster-bean, etc. The oat crop is usually grown pure but in northern Gujarat, rape is sown to form a subordinate mixture.

Cultivation

The land should be ploughed well or worked with a blade-harrow three or four times. There should be a gap of about a month between the harvesting of the previous crop and the sowing of oats to allow the stubble to decay well. However, if the oat crop is to be grown immediately after harvesting the previous crop, it would be desirable to remove the stubble to ensure a good germination of the seed.

The sowing season extends from October to December. For multicuts, the crop should be grown in

October. The seed is sown broadcast in the case of the fodder crop, and in rows 23-30 cm apart, if the crop is meant for grain. The seed-rate varies from 65-85 kg per ha, a larger quantity being used for sowings on poor soils. For fodder, a higher seed-rate is needed to ensure a dense stand of tall, thin-stalked plants.

The oat crop requires three to four irrigations. For maximum yields, it should be irrigated at 60 mm cumulative evaporation, as measured with the open-pan evaporimeter. The tillering stage is critical under moisture stress. Usually, no interculture is done.

When grown as a grain crop, the manure is generally applied to the previous crop in the rotation to prevent lodging. For the fodder crop, a fairly heavy basal application of organic manure supplemented by top-dressing with a nitrogenous fertiliser is practised to encourage the growth of the plant early.

The oat crop may advantageously be manured directly with a combination of nitrogenous and phosphatic fertilisers. The application of 40-100 kg of N per ha and 20-40 kg of P_2O_5/ha has been found to be very beneficial, depending on the crop in the preceding season, the length of the growing season, manuring, irrigation and the number of cuttings to be taken. The nitrogenous requirement of oats is more, following crops, such as maize, pearl-millet, sorghum, than after a fallow or legumes. Phosphatic fertilisers should be applied as a basal dose at the sowing time. Nitrogen may be applied in two equal doses at the sowing time and with the first irrigation or after the first cutting for fodder.

Harvesting, Threshing and Yield

The crop matures in about four-and-a-half to five months. For fodder, two to three cuttings are taken from January to March and the crop is allowed in some cases to set seed. For one cutting for fodder, the plants are harvested at 50 per cent flowering. For two cuttings, the first is taken at 65-70 days after sowing, and the second at flowering. After the first cutting the plants may be allowed to produce grains. However, this practice result in lower yields of grain, straw and dry matter than in those obtained from the first cut. Harvest the plants when the grains are mature, but the straw is still somewhat green; otherwise there is a loss of grain due to shedding. Threshing is done in the same manner as in the case of wheat and barley. With normal cultivation practices, the crop produces 500-600 quintals of green fodder and 2 to 3 quintals of grain per hectare.

Millets

Sorghum

Sorghum (*Sorghum vulgare Pers.*), popularly known as johar, is the most important food and fodder crop of dryland agriculture. The annual area under it ranges between 17 and 18 million hectares and the annual production between 8 and 10 million tonnes. Johar is mainly concentrated in the peninsular and central India. Maharashtra, Karnataka, Andhra Pradesh, Madhya Pradesh, Gujarat, Rajasthan, Uttar Pradesh (the Bundelkhand region) and Tamil Nadu are the major johar-growing states. Other states grow sorghum in small areas primarily for fodder.

The sorghum grain is used primarily as human food in various forms, such as roti or bakri (unleavened bread), or is cooked like rice. Sorghums are also malted, popped and several local preparations are made. Green and dried fodder is the most important roughage for feeding cattle throughout the country. The utilisation of grain sorghum as a cattle feed, poultry ration and other industrial uses is at present not very significant, although considerable scope exists.

Agro-climatic conditions

Medium and deep black soils are predominantly suitable for growing sorghums. Whereas the rabi sorghums are wholly confined to black cotton soils, the kharif sorghums are grown on light soils also on a limited scale. The sorghum belt receives an annual rainfall ranging from 400-1, 000 mm per annum, usually distributed between the last week of June and the first week of October in most parts of the country. Tamil Nadu and other parts of southern coastal Andhra Pradesh receive some rain during the north-easterly monsoon also. Within the limits of the season, temperature fluctuations are limited and rarely handicap crop growth, except when the rabi sowings of the crop are unduly delayed in the Deccan plateau.

Sorghums are grown during both kharif (July-November) and rabi (October-February) seasons, the rabi constituting 36-38% of the total acreage. The area under the rabi johar is concentrated in the Deccan plateau more or less a contiguous belt. In the states of Maharashtra and Karnataka, the rabi johar occupies about 55-60% of the total acreage, wheras in Andhra Pradesh, the distribution is 50-50 in both seasons. In the rest of the states the kharif season is more important. Only a very small area of this crop about 4-5 per cent, is grown under irrigation, usually during summer (January-April).

Cultivation practices

Sorghum cultivation still remains predominantly traditional in most parts of the country. The preparation of land with ploughs or blade harrows with least application of farmyard manure, line sowing with a seed-drill in rows 12-18 in apart, and interculturing with bullock-drawn implements continue to be practised and are very relevant even today. The factors which stand in the way of its production are the traditional late varieties which are drought-prone in years of submormal rainfall, the low crop stands and the absence of the use of fertilisers. Weed control may not be serious in black soils, but weeds are a limiting factor in kharif red soil. A two-year rotation of jowar-cotton is most common during kharif in the Deccan and central plateaux. The introduction of groundnut to make it a three-year rotation is more profitable. During rabi, the jowar-cotton, jowar-gram or jowar-jowar rotations are common.

Mixed cropping of jowar-arhar (tur) is most common. Mixtures with mung, urid, mash, cowpea and even with bajra and other cereals, vegetables, etc. during kharif are practised under different situations. During rabi the mixing of jowar and safflower is the most common practice. Even under rainfed conditions, the practice of sequence cropping is in vogue in areas of assured rainfall. Mung in kharif, followed by the rabi jowar, or the kharif jowar, followed by its ratoon or safflower, gram etc. are also becoming feasible, particularly with the introduction of short-duration varieties.

Harvesting and threshing are still carried out manually or with bullock power. Threshing by running a tractor or with power-operated threshers is practised on a very limited scale. The harvested grain is sun-dried and storage is conventional. The national average yields are still low and are around 500 kg/ha. However, the high-yielding hybrids and varieties yield 2, 00-3, 000 kg/ha under average growing conditions and the higher-order yields of 6, 000-7, 000 kg/ha are obtained under optimal conditions.

Forage Sorghums

Four improved varieties of forage sorghum, 'Pusa Chari' (IS 6090), 'Haryana Chari' (JS 73/53), 'SLOPE-44' and 'Meethi Sudan' (SSG 59-3), the first three suitable for the single cut and 'Meethi Sudan' for the cut have been released recently by the Central Variety-Release Committee for general cultivation. These fodder varieties now provide the necessary choice of forage varieties for different parts of the country.

Sowing

Sowing with the onset of the monsoon during kharif eliminates shootfly attack and gives high yields. This practice involves no money, but will require continued education to the farmers and propagation by extension agencies.

Plant Population

Unlike the locals, about 180, 000 to 200, 000 plants per hectare are essential if we are to realise the full potential of high-yielding varieties and hybrids. Such a population can be accomplished by using a seed-rate of 10 to 12 kg/ha. Seed costs are relatively low, and in the case of varieties they can be much lower. A plant at every 4 to 6 inches in rows spread 18 inches apart is essential.

Fertiliser Use

In addition to compost and farmyard manure, the applications of 30 to 40 kg of N and 30 to 40 kg of P_2O_5 per hectare in the form of inorganic fertilisers to high-yielding varieties would return (per kg of nitrogen) three times more grain yield than with the locals. In black soils, all nitrogen and P_2O_5 can be applied basally; in light-textured soils, nitrogen should be given in two doses, one-half basal and the other half at 30 to 40 days after sowing. Applying the available fertiliser to high-yielding varieties in itself results in the efficiency of fertiliser use. Nitrogen responses are optimum up to 80 to 100 kg of N/ha; this should be applied when the fertiliser supplies are not limited.

Pest control

Most of the sorghum pests can be avoided. Planting with the onset of monsoon avoids the shootfly. The en bloc coverage and the choice of similar-maturing types would avoid the midge. Both these aspects need only education and changes in the emphasis of extension efforts. If planting is necessary during the shootfly build-up, seed treatment with Carbofuran (5 parts (AI) for 100 parts of seed) gives effective control.

Stem-borer damage, which is not usually serious in most situations, can be checked by one or two applications of granules of 4 per cent Endosulfan, 4 per cent Carbaryl, or 2 per cent Lindane in the whorls.

A maturity-oriented policy of varietal spread can avoid the occurrence of the midge. The midge is effectively controlled by spraying the earheads before flowering (at about 50% panicle emergence) with Endosulfan (35% EC) one litre, Carbaryl (50% WP) 3 kg, Lindane (20% EC) 1 to 2 liters in 500 to 600 liters of water per hectare. The repeating of the spray after 4 to 8 days is desirable. Four per cent Endosulfan dust or 10 per cent Carbaryl dust or 10 per cent BHC dust at 20 kg/ha is also effective. Two applications will be necessary as in the case of sprays. The pre-emergence applications of Atrazine (0. 5 kg (AI) /ha), followed by a late-weeding control weeds.

Generally, the locals have shown a greater tolerance to shootfly than some of the high-yielding hybrids and varieties. The present efforts are to recommend high-yielding hybrids at the optimum time of sowing when there is no shootfly build-up. Treatments like Carbofuran seed treatment are recommended only in endemic areas or under conditions of pest build-up. As far as the midge is concerned, the locals showed greater susceptibility. The present efforts are to check it first through suitable varietal-planning, and resorting to protection only as the next step. Thus the emphasis on insect control is through pest avoidance, rather than through killing with chemicals.

Other Millets

Millets are warm-weather cereals with small grains and include six genera, i. e. Panicum, Setaria, Echinochloa, Pennisetum, Paspalum and Eleusine. The areas under different millets, as recorded in 1970-71 in India, were 12. 9 million hectares under pearl-millet, 2. 5 million hectares under finger-millet, and 4. 8 million hectares under other minor millets, with a total area of 20. 2 million hectares and a total production of 12. 1 tonnes of grain. The corresponding world acreage and production of millets (excluding sorghum and pearl-millet) are 35. 4 million hectares and 10. 3 million tonnes. Millets form a staple food in India, and in several developing countries of Africa, Near East and South Asia.

Barnyard millet

This plant is closely related to the weeds commonly found in dryland and sometimes in the rice fields. It is grown largely in Madhya Pradesh, Uttar Pradesh, Tamil Nadu, Andhra Pradesh, Karnataka, Maharashtra and Bihar. The yield of this crop is less than 400 kf/ha. This species is a quick-growing robust, tufted annual grass grown in kharif in most parts of the country and in rabi in the southern most districts of Tamil Nadu and in parts of the Deccan. It can be grown even on higher altitudes up to a height of 2, 000 m in the Himalayas. It is very drought resistant, but is also capable of withstanding water-logging.

Cultivation

Sowing with the onset of the monsoon during kharif eliminates shootfly attack and gives high yields. This practice involves no money, but will require continued education to the farmers and propagation by extension agencies.

Since this crop is raised in soils of marginal fertility, the sowings commence in June, with the onset of the monsoon. It is harvested by September during kharif and by January and February if sown during October-November as a rabi crop. Since it is cultivated in very light soils, the land is given only a limited preparatory cultivation comprising two ploughings. The seed is broadcast or drilled at 8-10 kg/ha. Transplanting is also done in some places in Maharashtra. It is raised as a pure crop or as a mixed crop with cotton, pigeon-pea or Dolichos or other short-duration pulses. The crop is seldom manured. However, in the Co-ordinated Trials during the past few years under the millet programme, yield levels up to 20 q/ha were recorded under better management in the same soil under barani (rainfed) conditions. When the crop is ripe, it is cut and stacked in the field for a week before threshing. The threshability of the grain is quite good.

Pearl millet

This crop is cultivated for grain as well as for fodder in the arid region of Africa and Asia and as a pasture in the USA. In India, the crop is grown over 12 million hectares, representing 30 per cent of the acreage of the world and 11 per cent of the total cereal production in India. It is grown all over India except Assam, north-eastern India, Himachal Pradesh, Jammu and Kashmir with Rajasthan, Maharashtra, Gujarat, Uttar Pradesh and Haryana accounting for nearly 5/6 of the total area.

Climate and soil

It is grown mostly during June to October and as a winter crop in March to June. During summer it is grown in Tamil Nadu, Karnataka, Punjab and Gujarat as an irrigated crop. Its range of adaptation is high under different day-lengths, temperatures and moisture stress. Most of the cultivars in India are relatively photo-insensitive as compared with the West African material which is highly photo-sensitive and very tall. It is grown on a wide range of soils, such as sandy loams of the Punjab and Uttar Pradesh and the light soils of Rajasthan and northern Gujarat, heavy clays of Andhra Pradesh, Tamil Nadu and very light soils as in Marathwada and the shallow black, red and light soils of the Deccan and southern India. It is eminently suited to light soils.

Cultivation

It is grown as a pure or mixed crop and is rotated with cotton, sorghum, niger, wheat and in rabi with pulses as a mixed crop. It is grown along-with a wide varieties of oilseeds and pulses. It fits into an intensive cropping pattern of 3 or 4 crops per year where irrigation facilities exist. The preparation of land is done on a very limited scale, since the traditional areas of cultivation are of light texture. The crop is sown immediately after the onset of the monsoon in early June in the Deccan and Rajasthan.

The locals are rarely manured under arid conditions, but the recent data on the fertiliser response of the new hybrids have shown that the economic optimum doses are 40 kg/ha in severe drought-prone areas, 80 kg N/ha in areas of limited to adequate moisture and 100 kg/ha in regions of adequate rainfall, with a high retentivity of moisture. The foliar application of half the recommended doses of nitrogen is suggested in the areas of severe drought. Under limited moisture, the yields of the hybrid varied from 1. 300 to 2, 400 kg/ha, depending on the nitrogen dose. The spacing is normally 60 cm between the rows and 15 to 20 cm within the rows. The available data on plant population and nitrogen application have shown that the row spacing can be reduced to 25 or 30 cm with 8 to 10 cm within the rows, particularly with the dwarf hybrids. With the new hybrids, it has been found that it is possible to maintain a plant population of 1, 75, 000/ha. Sowings are done normally in June-July for the kharif crop and in October for the rabi crop. The recommended seed-rate is 5 kg/ha. With some of the new dwarf hybrids, the plant population can be further increased with a spacing of 30 cm between the rows and 8 cm within the rows. The application of zinc sulphate at 2 kg/ha is found to be essential in light soils owing to the wide-spread deficiency of this element. The optimum date of sowing is found to be early July for a better establishment of the plants and for reducing the incidence of downey mildew.

Although this crop is traditionally grown mixed with others, there is a shift toward its cultivation as a pure crop consequent upon the availability of new hybrids. Interculture done in the early stage up to four weeks after germination is adequate to suppress the weeds. The crop matures in October, with 50 per cent flowering completed in 48-55 days and maturity in 88 to 96 days. The photo-insensitivity of the male sterile line has been responsible for this behaviour of the hybride. Owing to serious damage by birds, bristled ears are favoured. Plant-protection measures are essential against shootfly, white grub, and recently against downey mildew and ergot. Seed treatment against downey mildew and ergot with Thiram in the seedling stage and Ziran in the flowering stage against downey mildew and ergot respectively, and Endosulfan against the shoot-fly and white grub are effective. The crop is harvested close to the ground. However, the leaves of the hybrids are still green and can be used as fodder, when the locals are completely dry. The grain can be threshed after removing the ears from the harvested plants. The harvested crop is normally stacked for a few days. The dwarf hybrids are also suitable for harvesting with a combine.

Finger Millet

It is an important cereal in Karnataka. It is extensively grown in Karnataka, Tamil Nadu, Andhra Pradesh, Orissa, Bihar, Gujarat and Maharashtra and in the hilly regions of Uttar Pradesh and Himachal Pradesh, with a total area of 2. 5 million hectares and 2. 2 million tonnes of production.

It is grown in areas, with rainfall ranging from 50 to 100 cm and in irrigated areas. It is also raised as a summer crop and as a rabi crop in southern India, but mostly during kharif in northern India. The soils put under this crop are red loams, black and sandy loams in the South and alluvial soils in Gujarat, Uttar Pradesh and Bihar.

Cultivation

The irrigated crop is raised throughout the year in Karnataka, Tamil Nadu and Andhra Pradesh. The rainfed crop is sown during May or June and harvested in or is sown in September or October in rabi. The crop is manured liberally with 5 to 6 tonnes of farmyard manure and with split applications of nitrogen. The Co-ordinated Experiments have revealed that 40 kg N/ha under rainfed conditions and 160-200 kg N/ha with irrigation are economic optima, with responses of 6 to 28 kg of grain per kg of nitrogen.

Harvesting and Yield

The crop flowers in 60-80 days and matures in about 135 days, depending on the tract and the variety. The crop is cut close to the ground and ears are removed from the stacked sheaves. In the case of the irrigated crop, the ears are gathered as they mature, heaped for a few days to cure, and threshed. The average yield of the rain-fed crop ranges from 10 to 15 q/ha, whereas the irrigated crop yields up to 40 or 50 q/ha. The fodder yield ranges from 30-90 q/ha in the case of the early group and 95-150 q/ha in the case of the late group.

Italian Millet

Italian millet or fox-tailed millet is cultivated in tropical and temperate regions. The major regions of its cultivation in India are Karnataka, Andhra Pradesh, Madhya Pradesh and Uttar Pradesh. A majority of the varieties mature between 90 and 110 days. A new line, MS 1844/2, which will mature in 65-70 days, has been evolved recently. The crop can be grown even at an altitude of 2, 000 m. Light soils, including red loams, alluvials, and black cotton soils are all suitable for its cultivation. The crop is mostly grown under rainfed conditions and occasionally under irrigated conditions. It can be grown throughout the year in southern India. The rain-fed crop is mostly sown in June-July and harvested in September, whereas the irrigated crop is sown in February or March. In the southern districts of Tamil

Nadu, it can be sown in September and October as a rainfed crop and in March-April as an irrigated crop. It is generally sown mixed with other crops, but sometimes, it is raised as a pure crop. The common mixtures are cotton and finger-millet and a number of pulses. The crop is usually manured with 5-10 tonnes of farmyard manure per hectare. The agronomical data of the co-ordinated trials have shown that this crop can respond to nitrogen up to 60 kg/ha. The irrigated crop receives water at 10-day intervals. The crop is harvested, when the ears are dry, either by cutting the whole plant or the ears separately, threshing is easy and is normally done with a stone-roller.

Yield

The yield of the rainfed crops ranges from 500-1000 kg/ha and that of straw from 1, 000-2, 000 kg/ha. The irrigated crops yields 1, 000-1, 600 kg per ha, and fodder up to 40 q/ha. The crop flowers in 50-60 days and matures in 80-100 days. The grain has to be dehusked before cooking like rice or made into porridge. The straw can be fed to cattle without cutting.

Common Millet

It is a quick-maturing, highly drought-resistant grain crop, often grown during emergencies, famines and scarcities or when the proper season has passed for the sowing of the main food crops. Uttar Pradesh, Madhya Pradesh, Maharashtra, Andhra Pradesh, Karnataka and Tamil Nadu are the principle areas of its cultivation.

The crop is almost exclusively grown unirrigated as a late monsoon crop in poor soils, but in some parts of Tamil Nadu and Gujarat, it is cultivated as an irrigated crop sown in March to April. The seed is sown broadcast or drilled at 8 to 10 kg/ha and is covered with a brush harrow. The rows are about 25 cm apart. The crop seldom receives any manure or intercultivation. The irrigated crop is given one or two waterings. The crop matures in 90 to 100 days. The mature plants are uprooted or cut close to the ground and threshed immediately to avoid any shattering of the grains.

Yield

Four to eight quintals of grain is obtained from the rain-fed crop and 10-20 q/ha of it from the irrigated crop. The husk forms about 35 per cent of the grain yield and the husked grain is cooked and eaten or used for making chapatis. The yield of straw may vary from 10-25 q/ha. The straw is somewhat coarse but is used as fodder for cattle in an emergency. The irrigated crop produces good fodder.

Pulses

Pulses form an important part of Indian dietary. They are an important source of protein, are essential adjuncts to a predominantly cereal-based diet and enhance the biological value of the protein consumed. Being leguminous crops, possessing root nodules, they fix and utilise atmospheric nitrogen. They are thus not dependent on industrially fixed nitrogen, a process requiring energy, but add up to 30 kg of N/per ha to the soil and improve its fertility. The annual area under pulse crops in India is about 22 million hectares and the production is about 11-12 million tonnes of grain.

Black-gram

Black-gram (urd, mash) is a highly prized pulse, very rich in phosphoric acid. It is used, more or less, in the same way as green-gram (mung). It is also used in preparing papad (a kind of wafer) and barian (spiced balls of ground dal) and together with rice in preparing dosa and idli, popular breakfast dishes.

Mash is grown all over India, mainly in the monsoon season, the main areas of production being Madhya Pradesh, Uttar Pradesh, Punjab, Maharashtra, West Bengal, Andhra Pradesh and Karnataka. Annually, a little less than one million tonnes of grain is produced from an area of about 2. 5 million hectares.

Soil and climatic requirements

Though blackgram can be grown in a variety of soils ranging from sandy soils to heavy cotton soils, the most ideal soil is a well-drained loam with pH ranging from 4. 7 to 7. 5. On the other hand, blackgram cannot be grown in alkaline and saline soils. It is basically a warm-weather crop. In northern parts of the country, where the temperatures during winter are quite low, it is also grown during winter. In the central and southern states where there is not much variation in the climate, it is cultivated both during winter and rainy seasons.

Tillage

While preparing the land, the first operation should be done by a soil-turning plough and the subsequent 2 ploughings should be done by a country plough or cultivator. Where it is to be grown mixed with other crops, the soil preparation should be done according to the requirement of the main crop. Removal of the stubble of the previous crop overcomes the danger of termite attack.

Among the varieties mentioned, 'T 9' is early-maturing (75-80 days), dwarf, photo-insensitive and suitable for planting in all the seasons-spring summer and rainy seasons, throughout the country. It also fits well in different multiple-cropping systems. 'Pusa sel 1' is high-yielding, medium in maturity (85-90 days) and fairly resistant to viral diseases.

Application of fertilisers

A starter dose of 10 to 15 kg nitrogen and 25 to 40 kg P2O5 per hectare is adequate for blackgram. Application of 70 to 80 kg diammonium phosphate (DAP) per hectare is most convenient, as it meets adequately the total requirement of nitrogen and phosphorus. Molybdenum also helps in increasing crop production.

Weed control

When sown as a sole crop, 1 or 2 weedings are needed in the initial stages to keep the field free from weeds. However, 1 weeding 30-40 days after sowing is found to be enough. The pre-emergence application of Lasso @ 1. 5kg ai/ha keeps the field from weeds for the first 50 days and thereafter the crop smothers the weeds.

Irrigation

Blackgram is grown purely as a rainfed crop during kharif and on residual moisture in rice fallows during rabi. In case there is a long dry spell, the crop is benefited if 1 irrigation is given. Similarly, during the winter season, if irrigation facilities are available, 1 light irrigation at podding/seed-filling stage is beneficial to the crop. Excess water at any stage affects the crop adversely.

Chickpea

Soil or climatic requirements

Chickpea thrives best on well-drained, coarse-textured/ loamy soils, but also grows fairly well in black cotton solis or medium clay soils. It can also grow successfully in poor soils which cannot sustain other major cereals or commercial crops. It can stand a wide range of soil pH (5. 5-8. 6) but is very sensitive to saline and alkali soils. Waterlogging also adversely affects this crop. Experience has shown that well-drained clay loams with adequate moisture-retention capacity are the best for getting higher yields in chickpea

Tillage

In large areas, chickpea is sown after the monsoon on fallow lands. The soil moisture is conserved through appropriate field operations, like a deep ploughing followed by 2 or 3 harrowings and removal of all weeds and stubble of the preceding crop. Chickpea does not require fine tilling but the soil should be well-opened and loosened for better aeration. This will have a remarkable effect on the reduction of wilt incidence and better yield performance. Deep tillage up to a depth of 25 cm during July and August will have substantial effect on better root development, moisture conversation and crop yields

Seeding technique

The best time for sowing is during mid-October with the onset of cool weather when the daily average temperature is around 25°C. If winter is delayed, sowing can also be delayed with advantage up to the first week of November in the northern region; however, in the central, western and southern regions, sowing need to be completed by the middle of October. Earlier sown crop suffers form wilt disease owing to higher temperatures at sowing time. Early sowing therefore, gives poor yield. The optimum time for sowing Kabuli type is the last week of October. As an unirrigated crop chickpea is normally grown after monsoon fallow or after the harvest of kharif crop utilising the residual moisture left in the soil. It is taken either as a sole crop or as an intercrop with wheat, barley or mustard. Depending upon the seed size the recommended seed rates are 60-75 kg/ha and 80-100 kg/ha for deshi and Kabuli types, respectively. Chickpea may be sown in rows keeping the row-to-row spacing between 25 and 30 cm. To ensure seed placement in moist zone, pora may be attached to the country plough.

If the crop is planted under dryland condition, the seeding at a depth of 10-15 cm is beneficial on two accounts. First the seedling takes moisture from deep strata of the soil and secondly the root system escapes the zone where fungus causing wilt is concentrated. Under irrigated conditions, seeding at a depth of 5 cm may be maintained, though deeper sowing is still preferred.

Application of fertilisers

Application of 10 kg nitrogen along with 20-30 kg P_2O_5/ ha, generally increases the yield of the unirrigated crop over a wide range of conditions. When sown after kharif fallow nitrogen may not be applied. When chickpea is sown under irrigaitions, the dose of phosphorous should be 40-50 kg of P_2O_5/ ha. If chickpea succeeds crops, like maize or sorghum, nitrogen should be applied @ 20kg/ ha. Phosphorous may be given through single superphosphate, which is a better carrier. Response of the irrigated crop to 33. 6 and 67. 2 kg P_2O_5/ha is about 2. 6 and 4. 5 q/ha of grain respectively with an average yield of 11. 5 q/ha in the control (no manure) plots. To obtain a good response to applied fertilisers, moisture conversation practices may be adopted.

Weed control

One hand-weeding 45 days after sowing has been found

to be the best. Where the land is infested with a variety of weeds, 2 weedings at 30 and 60 days after sowing could be done depending upon the conditions of the field. Where chickpea is grown on large areas, use of herbicides could be made. Basalin @ 1. 0 kg ai/ha gives results. The alternative herbicides are Stomp (0. 75 kg ai/ha) or Ronstor (1. 5 kg ai/ha).

Chickling pea

This crop is chiefly cultivated for fodder. Poor people use the grain as pulse, particularly in times of security. The unripe pods are occasionally used as a green vegetable. The grain and other plant parts contain neurotoxin, BOAA, which has been shown to be the inciting agent of lathyrism, a paralysis of the lower limbs, endemic in some regions of Madhya Pradesh, Bihar and Uttar Pradesh. Ill-effects are noticed when lathyrus forms a significant part of the diet over a considerable period. Steeping the pulse overnight in water and rejecting the water effectively detoxifies the pulse. Parboiling has also the same effect. A variety ('P24') has become available. It has a very low BOAA content and, therefore, is not likely to cause lathyrism.

Soil and season

It is a cold weather drop which thrives best on low-lying water-retentive heavy soils. Deep black soils produce excellent crops. The crop is not injured much by the cracking of black soils. Being hardy, it can be grown on lands which are suited to no other crop. It is grown either as the sole crop of the year or as a second crop after paddy, being often sown as an utera crop in standing paddy. Sown in September-October, it is harvested in February or March.

Cultivation

The crop is frequently grown on land which might have remained submerged in water for a part of the kharif season. Two or three ploughings are soon given after the rains to secure a fairly clean seedbed. The seed is sown broadcast, in rows, using 10 to 35 kg per hectare, the broadcast crop requiring higher seed rate. When growing utera after paddy, the seed is broadcast in the standing cereal crop a few days before harvesting. One or two hand-weedings are beneficial.

The crop, when ready, is cut with a sickle, dried on the threshing floor for a week or so, and threshed by trampling it with bullocks. The average yield of grain per hectare is 300 to 400 kg. Yields of 600 to 800 kg are not uncommon.

Cowpea

Cowpea is used as a fodder, vegetable, a pulse and a green manure crop. It is grown all over India, more particularly in the central and peninsular regions, though on a limited area. It has considerable promise as an alternative pulse crop in the dry land farming.

Soil and climatic requirements

Cowpea is a tropical crop and prefers warm and humid season and grows well on a wide range of soils including poor soils with low pH. However, saline and alkali soils are not suitable. Although it is mostly grown as a rainfed crop, it is capable of withstanding high rainfall, but is less tolerant to waterlogging.

Seeding technique

Depending upon whether the cultivar is compact, semi-spreading or spreading, seeds are sown @ 10 to 40 kg/ha and the rows are spaced at 30-90 cm. For grain types, the row-to-row distance should be maintained at 30 cm, and the plant-to-plant distance between 5 and 10 cm depending upon the plant type. The seed rate should be around 25 to 30 kg/ha for a sole crop.

Application of fertilisers

Starter dose of nitrogen @ 15 kg/ha along with 30 to 40 kg of P2O5/ha is considered, optimum. If the crop is grown consequent to a cereal or a commercial crop, the cowpea crop need not be fertilised as residual fertility would be enough to sustain the crop. Care should be taken that phosphorous is placed 5 cm below or away from the seed.

Plant protection

Cowpea is attacked by pests, like leaf-miner, hairy-caterpillar, jassids, aphids, white-fly and galerucid beetles; and diseases, like Cercospora leafspot, bacterial leaf-spot, cowpea tip-necrosis, cowpea bean-mosiac and cowpea banding-mosiac virus etc.

Green gram

Green gram or mung about 2. 5 million hectares, is spread all over the country, with a production of about 0. 8 million tonnes. The important states producing this pulse are Madhya Pradesh, Maharashtra, Uttar Pradesh, Punjab, Andhra Pradesh, Rajasthan, Karnataka, and Tamil Nadu. The grains (whole or split) are used as pulse or made into flour, and into straw and husk as fodder for cattle. Unlike other pulses, it does not produce a heaviness or flatulence. Grains are also eaten whole (after germinating them), parched, salted, with sugar or boiled with condiments.

Kharif greengram

The kharif crop is grown both as sole and as intercrop. Since it is largely grown as a rainfed crop, the sowing time usually depends on the onset of the monsoon in the different regions. Normally the sowing is done from mid-June to second week of July, which is considered optimum. Delayed sowings would result in progressive decline in yields.

Soil and climatic requirements. Greengram usually grows on a wide range of soils varying from sandy loam to alluvial, red laterite or black cotton soils. However, it has been observed that crop thrives best on lighter soils with good drainage. In India, it is grown from sea level up to an altitude of 2, 000 m largely as a dryland crop. Although fairly drought- resistant, the crop is susceptible to waterlogging and frost.

Seeding technique. Row spacing of 30 cm is considered optimum, although 45 cm row spacing is adopted in case of spreading types. By and large, a spacing of 25-30 cm between rows depending on the canopy development of the variety and date of sowing is adequate. The plant-to-plant distance should be maintained at 5 cm under irrigated and assured rainfall conditions and 10-15 cm under rainfed conditions. A seed rate of 15-20 kg/ha depending on the seed size of the variety is optimum. For very bold-seeded types, a seed rate as high as 30 kg/ha may be required.

Application of fertilisers. The response to phosphorous is highest on red and laterite soils. Application of P2O5 @ 30-40 kg/ha along with a starter dose of 10-15 kg nitrogen is adequate. Phosphorous application has always significant effect in increasing the yields. Seeds should be treated with an efficient Rhizobium culture for obtaining higher yields. Rhizobial inoculation may reduce the nitrogen requirement of the crop.

Weed control. The highest crop yield was obtained when weeds were removed 35 days after sowing. Any further delay in weed removal results in a corresponding decrease in yield. A maximum of 2 hand-weedings in the initial stages of crop growth up to 30-35 days, is adequate to take care of the weed problem. However, whenever labour is in short supply or the rainfall pattern does not allow early hand-weeding, herbicides need to be used. Pre-emergence application of Lasso or Tok E-25 @ 2kg ai/ ha in 1, 000 litres of water ensures complete weed control.

Irrigation. Greengram does not require any irrigation if the monsoon rainfall is well distributed. However, in case of drought conditions prevailing during the growth stages due to failure of rainfall for longer spells, the crop gets benefited by irrigation. One irrigation under drought situation at flowering stage particularly in sandy loam areas is recommended.

Summer greengram

Soil and tillage. Summer greengram can be grown after harvesting of wheat crop without any preparatory tillage. It could be seeded by opening a small furrow in between the rows of wheat stubble and irrigated immediately thereafter. These operations would require minimum tillage as well as time and operational costs. However, in order to obtain a good crop, a very heavy pre-sowing irrigation (double palewa) may be given and the field ploughed twice with harrow to give a good tilth.

Seeding technique. A distance of 25 cm between rows and 5 cm between plants is optimum. The highest yield of summer mungbean can be achieved when seed rate is about 25 kg/ha or even more.

Applicaton of fertilisers. A starter dose of 10 kg of nitrogen/ha along with 40 kg P_2O_5/ha is optimum for summer greengram. In a 3-crop sequence of maize-wheat-summer greengram, the greengram need not be given any nitrogenous or phosphatic fertiliser, if the previous 2 cereal crops had received the recommended doses of nitrogen and phosphorous. The fertilisers may be drilled through pora in furrows drawn 25-30 cm apart. Fertilisers may be drilled along with the seed or placed 5-6 cm below the seed through pora It is also necessary to treat the seed with an efficient Rhizobium culture.

Weed control. Two hand-weedings, the first 25 days after sowing and the second 45 days after sowing, are adequate to check weed infestation. Subsequently, greengram grows rapidly and the weeds are smothered. Alternatively, pre-emergence weedicides, like Tok E-25 or Lasso 1 litre/1, 000 litres of water may be sprayed in a hectare, just after sowing. However, weedicides control only broad-leaved weeds whereas motha (Cyperus rotundus) is a major problem in the summer season. Therefore, one handweeding, preferably before the first irrigation will take care of this problem.

Irrigation. The number of irrigations and their time of application vary according to seasonal conditions. At least 3 irrigations, the first at preflowering stage (20-25 days), the second at flowering (25-40 days) and the third at grain-filling stage, are necessary. Pre-sowing irrigation is a must to ensure adequate soil moisture for germination. The availability of water is generally scarce in the canals during the summer months but there is a great scope of growing summer mungbean around tubewells. There are about 6. 14, 1. 24, 0. 11 and 1. 86 tubewells in Uttar Pradesh, Bihar, Rajasthan and Haryana respectively to support summer cultivation of greengram.

Moth bean

Moth is the most drough-resistant of the kharif pulses,

and is largely grown as a dry crop, either alone or subsidiary to millets or cotton. It is also grown mixed with irrigated fodder jowar and bajra. On account of its material-like habit of growth, it is highly useful against wind-erosion in sandy areas. Occasionally, it is used as green-manure, ripe pods are consumed as a green vegetable. The whole grain is used as pulse in the main producing regions. Green as well as dry plants make good fodder. Rajasthan accounts for over four-fifths of the area (1. 7 million hectares) and three-fourths of the annual production (0. 2 million tonnes). Uttar Pradesh, Punjab and Maharashtra are the other moth-producing tracts.

Seeding technique

Mothbean requires a minimum preparation of land and can grow on all types of soil. It is primarily a kharif crop and does well on dry sandy soils with poor water retention. It does require good drainage. As a sloe crop, mothbean may be sown in rows 40-80 cm apart at shallow depth. The seed rate is 10-15 kg/ha for the sole crop and somewhat less when grown as an intercrop. One weeding may be done during the early growth stages to reduce competition from weeds. Application of 10 kg of nitrogen and 20 kg P_2O_5/ha are adequate for the crop.

Soil and climate

This pulse is largely cultivated on well-drained but poor soils, ranging from very light alluviums to heavy loams. In the Deccan, it is grown on light, stony, more or less, upland soils. It prefers a well-distributed moderate rainfall. Heavy rain is very detrimental, because it results in profuse vegetative growth.

Cultivation

The cultivation of this pulse is similar to that of green-gram (mung) and black-gram. The preparation of soil is very haphazard. Sowing starts with the monsoon rains in June or July. The seed is sown broadcast at the rate of 25-40 kg per hectare when grown alone, or at the rate of 5-25 kg when mixed with other crops. No weeding or interculture is given. The crop is ready for harvesting in October-November. The ripe plants are cut with a sickle, dried for a week or so and threshed by trampling with bullocks and are winnowed as usual. The average yield is about 2-5 quintals of grain per hectare.

Peas

Peas (*Pisum sativum L.*). The major part of the area under this crop is in the Indo-Gangetic plains, Uttar Pradesh, accounting for 0. 5 million hectares and 0. 47 million tonnes (out of 0. 65 million hectares, with a production of 0. 5 million tonnes) in 1974-75. It is a high-yielding pulse crop; unripe pods are used as a green vegetable and the broken parts are used as a cattle feed.

Soil and Climate

This pulse is similar to gram and lentil in it requirements, except that it does better in loamy soils. It cannot be grown in alkaline soils. It is grown as a cold-weather, irrigated crop. The sowing season extends from October to December. It is generally cultivated after cotton in Uttar Pradesh.

Cultivation

This pulse resembles gram a great deal in respect of cultivation except that a finer seedbed is needed. Seed is sown broadcast, or in rows 30-45 cm apart. The seed-rate is about 100 kg per ha. The crop generally receives no manure, weeding or interculture, but has been shown to respond to these inputs. Powdery mildew often causes appreciable loss in yield. This disease can be controlled economically by spraying it once or twice with wettable sulpher. The crop takes 130-140 days from sowing to maturity. The crop which receives irrigation remains protected against frost to a great extent.

Pigeon-pea

Pigeon-pea (*Cajanus cajan* (L.) Millsp.). Pigeon-pea or red-gram (arhar) is extensively used as dal; its green podes may be used as a vegetable. The green leaves and tops of plants are fed to animals or are utilised as green manure. The husks of pods or seeds with parts of the kernels (a byproduct of dal preparation) constitute a valuable cattle-feed. Dry stalks obtained after threshing are used for basket-making or as fuel or as thatching material. Being deep-rooted, it is also planted as a soil renovator to break up the hard subsoil and as a hedge to check erosion. The heavy shedding of leaves adds considerable organic matter to the soil. It is also often grown as a cover crop in plantations.

Soil and climatic requirements

Pigeonpea is grown throughout the tropical, subtropical and warmer regions of the world between 30° north and 35° south latitude. As it prefers warm temperatures, pigeonpea is grown in the plains of India during different seasons in different parts of the country. However, major cultivation is confined to the kharif season. Though it can withstand high temperaratures, low temperatures are unfavourable for its growth and development.

Further, it is susceptible to frost and even one night's frost can cause heavy losses. As such in northern parts of the country, pigeonpea is grown as a summer or kharif crop, whereas in central and peninsular regions it can be grown throughout the year. Pigeonpea can be grown on a wide range of soils varying from loam to very heavy clay soils. It thrives well on red soils as well as black cotton soils of Maharashtra and Madhya Pradesh. Pigeonpea with its deep root system is fairly drought-tolerant but cannot thrive in flooded fields; even temporary standing water causes heavy mortality, especially at a seedling stage.

Seeding technique

Pigeonpea is sensitive to excessive moisture, and proper land levelling and good seed-bed preparation are necessary. Two or three ploughings folllowed by planking are adequate. In rainfed areas sowing is done with the onset of the monsoon in kharif. However, sowings from the first week of June to the second week of July are considered best. Generally, delay in sowing beyond 15 July not only reduces the plant height and total dry matter but also decreases the number of pods per plant. Therefore, it is suggested that the sowing be organised either with pre-sowing irrigation or immediately with the onset of the monsoon but never beyond the last week of July.

The traditional long-duration types are preferred for mixed cropping with kharif crops, like sorghum, pearlmillet, maize, rice, fingermillet, and other pulse crops. However, if pigeonpea is grown in a double-cropping system either in rainfed areas or with limited irrigation, one should choose the earliest maturing varieties. Extra-early varieties (120-140 days) can mature in time even under delayed sowing and thus vacate the fields for timely sowing of wheat/mustard, etc. Performance of extra-early varieties, however, is poor in high-rainfall areas. Cloudy weather and heavy rains at the time of flowering and fruiting result in poor pod setting and filling together with increased damage by the pod-borers. With long duration varieties, spacing of 75 cm between rows and 30 cm between plants, and with the short-duration varieties spacing of 30 cm betweem rows and 20 cm between plants is optimum. However, under dryland/rainfed conditions, row distance could be widened up to 90 cm particularly with the tall, spreading, long-duration varieties; plant-to-plant distance within the rows should be maintained at 30 cm. The depth of sowing should be maintained at 4-6 cm.

Application of fertilisers

A starter dose of 15-20 kg nitrogen/ha needs to be given to ensure quick, early growth of the crops. Such a starter dose however, may not be necessary, if the previous crop was either a legume, or a non-legume crop which had received a heavy dose of nitrogen. For early and effective nodulation seed treatment with Rhizobium is recommended. Application of phosphatic fertilisers @ 40-50 kg P2O5/ha by placing it 6-10 cm below the recommended. In case of mixed cropping, an additional dose of 20 kg P2O5/ha at the time of sowing will be beneficial to both the constituent crops.

Weed control

Pigeonpea is a slow-growing crop during the first 6-8 weeks, and 2 harrowings during this period would be adequate to check weed growth. However, since the major area of pigeonpea is grown during the monsoon season, incessant rains may sometimes interfere with harrowings. Under such conditions, it would be necessary to use herbicides for control of weeds.

Irrigation

Normally the crop is grown under dryland conditons and is fairly drought-tolerant. However, the crop does respond to irrigation. Two irrigations, one at flower-initiation stage, and other a pod-filling stage, are recommended. In areas where irrigation resources are available, the approach would be to increase the area under irrigated pigeonpea. Water may be made available to the farmers from mid-May or June for undertaking sowing operations. The subsequent irrigations may be required in September/October and December.

Starch Crops

Potato

The potato, a native of South America, occupies the largest area under any single vegetable in the world. It has a special value as food. The consumption of potato per head is very low in India compared with that in Western countries. Potato belongs to the family Solanaceae, genus Solanum and species tuberosum.

The potato having continued under cultivation in India since the seventeenth century, there are some old varieties, like 'Satha', 'Gola', 'Phulwa', 'Up-to-date', 'Darjeeling Red Round', 'Craig's Defiance', 'Great Scot' and 'President', which are still cultivation.

Cultivation

Potato tubers are planted either whole or cut into pieces. Diseases-free certified seed should be sown. The seed weight varies from 800 to 1, 500 kg per hectare according to the size of the seed-tuber. In the plains of northern India, the sowing time is from the middle of September to the beginning of January. Two successive crops can be raised on the same land. In the Nilgiri Hills, three crops are raised in succession in April, August and January. In the hilly regions of the north, the crop is sown in March-April. A spacing of 45 to 60 cm X 15 to 25 cm, according to the variety and the size of tuber is adopted.

Potato needs irrigation at frequent intervals, depending upon the soil and climatic conditions. Usually, 6 irrigations are sufficient. Chemicals, such as Linurion or Simazine, at 0. 5 kg per hectare applied as a pre-emergence spray are effective in controlling weeds. Potato is harvested from the time it is of sufficient size until the vines are fully ripe. The average yield varies from 20 to 30 tonnes per hectare. The best method of storing potato is in cold stores at 2. 2°C to 3. 3°C and at 75 to 80% relative humidity. When stored in a room, good ventilation should be provided.

Aphids, jassids and cutworms are the most important insect pest of potato. They can be controlled by spraying Malathion, Folidol, DDT and Endrin. Important diseases attacking potato are early, and late blights and virus diseases, A, X and Y and leaf-roll. Charcoal-rot is the most serious storage diseases. The seed-plot technique consists of the sowing of tubers before the middle of October, the killing of haulms by the end of December before the aphid population builds up, and harvesting by the end of February or in the beginning of March. Two inspections made to rogue out any diseased or off-type plants, along with the adoption of the above techniques will produce disease-free seed.

5

Cultivation of Plantation Crops

Crops may be called plantation crops because of their association with a specific type of farming economy. Most of these involve a large landowner, raising crops with economic value rather than for substistence, with a number of employees carrying out the work. Often it referred to crops newly introduced to a region. In past times it has been associated with slavery, indentured labour, and other economic models of high inequity. However, arable and dairy farming are both usually excluded from such definitions. A comparable economic structure in antiquity was the latifundia that produced commercial quantities of olive oil or wine, for export. Plantings of a number of trees or shrubs grown for food or beverage, including tea, coffee, cacao and rubber are generally called plantations. Some spice and high value crops grown from permanent perennial stock, such as black pepper may also be so called.

Tea

Tea (*Camellia spp.*) is an important beverage and the world drinks more of it than any other beverage. It is made from the tender or young leaves and unopened buds of the evergreen tea-plant, popular as a 'healthful herb'. The tea plant is in the polymorphic species *Camellia sinensis* Kuntz, but recent findings show that this plant of commerce is derived from more than one species. Two distinct varieties of tea-plant are generally recognised, the small-leaved China (sinensis) and the large-leaved Assam (assamica) which have been raised to a specific rank by a well-known tea botanist. Careful field observations, however, reveal that more than one or two species are involved in the evolution of the present-day tea-plant of commerce. Considerable interspecific hybridisation has taken place in nature. Thus, the taxonomy of the tea-complex is confounded.

The tea-plant, in the natural state, grows into a small or medium-sized tree, but in commercial plantations it is pruned and trained to form a many-branched low bush and is encouraged to produce vigorous vegetative growth by adopting an appropriate schedule of fertiliser applications.

The important tree growing countries are India, Sri Lanka and East Africa; Japan and Indonesia also produce sizeable quantities of tea. It is also grown in Bangladesh, China, Georgia, Argentina and some other countries. In India, the crop is grown in Assam, West Bengal, Kerala, Karnataka and Tamil Nadu, and to some extent in Tripura and Himachal Pradesh. The total area under tea in our country is about 3,58,000 hectares and over 468 million kilograms of product is obtained annually. Over one million workers are employed by the tea industries.

Yields have steadily increased during two decades. Tremendous progress has been made on the selection of clones and those with high yields and improved quality have been released to the industry. The tea planters are encouraged to uproot poor section of their plantation and replant them with improved planting material.

Climate

Tea is grown on plantation scale in many areas of the world in warm and moist climate. These range from Georgia (USSR) to Corrientes (Argentina). It is grown from sea level to about 2460 m. A well distributed rainfall is essential for its cultivation. The rainfall varies from 125 cm to 750 cm in the tea-growing regions in this country.

Soil

Tea is planted on soils of widely different geological origin and of almost all physical types. A well drained, deep and friable loam or forest land rich in organic matter is ideal. Tea soils are acidic (pH 4 to 6), low in calcium and generally rich in iron and manganese.

Cultivation

Tea is propagated from seed, but high yielding clonal material has also become popular. The plants are initially raised in nurseries to produce healthy plants. 1 to 1 and 1/2 year old nursery seedlings are used for planting in the field. The vegetative propagation of selected tea-plant cuttings, each with a leaf and an internode has been standardised and progressive planters now raise colonial nurseries on a large scale.

The area intended for planting is first cleared of forest growth and adequate steps are taken to prevent soilerosion. Planting lines are traced (as at present on contour on hill slopes) and pits of a convenient size (generally 30-45 cm deep and 24 cm wide) are dug. The normal accepted spacing is 120 X 75 cm, accommodating about 10,000 plants in a hectare. The soil is heavily mulched during the initial one or two years.

In most of the tea areas in our country, a system of shading the tea plants is practised. In southern India, the silver oak (*Grevilla robusta*) is a shade tree of choice (planted 600 by 600 cm, later thinned to 600 cm by 1200 cm or 1260 cm by 1200 cm as required). Other trees used for shading are *Erythrina lithosperma* (dadap), *Acacia* and *Albizia. Albizia* species are used in tea gardens in northern India.

Weeds compete with the tea plant for nutrients, and weed control is important in tea plantations. In recent years, the control of weeds with chemicals has gained popularity. Grasses and broad-leaf dicot weeds are kept under check. Pests and diseases are effectively controlled with chemicals.

A manuring scheme is followed from the nursery stage upto maturity. A special 1:1:1 NPK soluble fertiliser mixture is used in the nursery; a 4:3:3 NPK mixture in the first 3 years in the clearing and thereafter the dosage of the fertiliser applied is varied, depending on the type of tea, its performance, soil fertility and the yield potential of the crop. On an average 10 kg of nitrogen is applied for every 100 kg of the crop.

Tea readily responds to the application of nitrogen and potassium. Potassium is applied at the rate of 40 to 50 percent of the amount of nitrogen applied, depending on the soil type and weather conditions. In southern India tea responds to nitrogen upto 300 kg per hectare applied to some high-yielding varieties of tea, whereas in northern India, the application of nitrogen is limited to about 120 kg whereas 30 to 40 kg of phosphate per hectare is adequate.

Zinc deficiency is a limiting factor in crop production in most areas and a schedule of foliar application of zinc sulphate usually 11 kg per hectare in alternate years is recommended.

Training, Pruning and Plucking

The tea-plant is initially trained into a small bush by centring low within a few months of planting, by removing the central leader stem in order to encourage development of lateral branches. The lateral branches are cut to a convenient height of 40 to 50 cm and the growth above this is periodically cut. Thus a small, compact bush is formed. New shoots are allowed to grow unhampered and these shoots are tipped, leaving a growth of 20 to 30 cm above the pruning cut, depending on that kind of plant. The crop is then harvested at regular intervals.

In southern India, the pruning cycle extends over a period of 4,5 or 6 years depending on the elevation and growth. In some caes, a 'skiff' is given at a cinvenient height and the pruning cycle extended for 2 or 3 years. Pruning and skiffing are done periodically to keep the height of the bush at a cinvenient level for the pluckers to operate and to encourage vegetative growth. Annual pruning is a practice in northern India (Assam), but even there the present trend is toward an extended pruning cycle. After a series of pruning cycle, the bushes are rejuvenated by hard pruning, removing all cankered and diseased portions of the stem.

In northern India, the economic life of the tea bush is generally 40 to 50 years and therefore 2 to 2 and a 1/2 percent of the area is uprooted and replanted every year. In southern, such regular uprooting and replanting is not practised, because economic yields are obtained from sections which are 80 years or more old. Usually plucking is restricted to the terminal bud and two expanded leaves, or to the bud and three succulent leaves, known as fine or medium plucking. Coarse plucking includes extra leaves. In northern India the tea bush is plucked at weekly intervals from April to December and there is a dormant season during winter. In southern India the crop is harvested during the cold and dry months (December to March).

Tea Manufacture

The shoots may be prepared in several ways to make black or green tea. To make the former, the leaf is usually withered and rolled or distorted by rolling in the conventional tea rollers, or is passed through tea machines, the 'rotorvane' and CTC rollers which exert drastic action on the withered leaves. The juice of the leaf cells is exposed in the air to oxidise (fermentation) when important changes take place. When the optimum fermentation is reached, the action of enzyme is arrested by drying the fermented leaf in a current of hot air in suitable tea-dryers. In the production of green tea, the shoots are steamed as soon as possible after plucking and the leaf oxidised is soon destroyed so that fermentation does not take place on rolling.

The processed tea is passed through a series of meshes, sorting out leaf grades and dust. The tea leaves are sorted into different grades, for example, pekoe, Orange pekoe, Flowery pekoe, Broken Orange pekoe and Fannings. Tea intended for export is packed in plywood chests and shipped directly to various centres or sold through auctions at centres like Calcutta and Cochin. Instant tea, soluble in water, tea bag and ready mix tea are new developments.

Tea research is carried out in the country at two centres: Tocklai Experimental Station of the tea research association, Jorhat (Assam) and the Tea Research Station of the United Planters Association of Southern India (UPASI) Cinchona P.o., Coimbatore district, Tamil Nadu. Certain adhoc schemes on tea research are carried out at centres such as the Central Food Technological Research Institute at Mysore and the Department of Applied Chemistry, Calcutta university. The Tocklai Experimental Station publishes its findings in its annual reports, bulletin the Tea Encyclopedia and in the journal Two and a Bud. The UPASI tea Research Station publishes the results of research in its annual reports, bulletins Handbook of tea Culture and in the Planter's Chronicle.

Coffee

Coffee is cultivated commercially in the four southern states of Karnataka, Tamil Nadu, Kerala and Andhra Pradesh. It is also grown on a limited scale in some non-traditional areas of Orissa, West Bengal, Assam and Madhya Pradesh

Arabica (*Coffee arabica*) and robusta (*Coffee canephora*) are the two principal economic species extensively cultivated in India. While 90,211 hectare area (57.98%) is under arabica, 65,365 hectares (42.02%) are under robusta. India produces 85,000 to 100,000 tonnes of coffee and earns considerable foreign exchange. Coffee cultivation is confined to the hilly areas of the Western and Eastern Ghats. The annual rainfall ranges from 1250 to 3000 mm.

A major portion of the area is under the south-westerly monsoon and only a small area is under north-easterly monsoon. This plant grows well at temperatures between 12 and 36 degrees Celsius. Elevation above the sea level influences the quality of the coffee. Coffee arabica grows well at elevations between 900 and 1200 metres; Coffee canephora or robusta grows luxuriantly at the lower elevations (about 150 metres). Arabicas are more shade-loving than robustas under conditions in south India.

Soil

Coffee soils in India belong to the red and lateritic soils. The soils differ in texture from sandy loam to clayey loam. Red ferruginous clays occur in tracts e.g. Bababudangiris in Karnataka and the Nilgiris in Tamil Nadu. They have good aggregating ability and are generally well-drained. They are rich in organic matter and are acidic to neutral. The total soluble salts are well below the sensitivity limits. They are well stocked with potassium but low in available phosphorus. They are also poor in calcium, and magnesium. They respond well to liming, manuring and other soil management practices.

Climate

Climatic and environmental factors, like rainfall, temperature, elevation and aspect can influence the economic production of coffee much more than soil factors. Under conditions in southern India, summer temperatures combined with poor subsoil moisture can be a severe limiting factor, whereas at northern latitudes cold winter temperatures can be equally limiting.

Varieties

There are many commercially important varieties and selections of c.arabica grown in India. Their features are described below:

Old Chiks. The earliest strain of C.Arabica to come under cultivation owes its name to Chikmagalur where it was initially grown. It is a tree of coarse growth, with branches having an upward turn and foliage of pale colour. Leaves are small and narrow, 15 cm in length. Fruits are red, beans are bold. The produce is reputed for its high liquoring quality. Heavy crops in alternate years appear to be a regular feature. The plant is highly susceptible to leaf rust.

Coorgs. is thought to have originated in Nalakanand area of Coorg, in about 1870. It has a graceful drooping margin. The fruits are bold, oblong and red. It is highly susceptible to leaf rust.

Kents. Developed by L.P.Kents of Doddangudda and Kalagnee estates in Mudigree zone was cultivated during 1918-1920. It is resistant to races II and VII of leaf rust. Leaves are larger than 'coorgs', dark green and thick. Apical leaves are dark bronze. Fruits are red. Beans are bold, oblong and larger than those of 'coorg'. Kents attracted the attention of the coffee growers in India. The strain was also introduced in certain African countries. This is high yielding, capable of withstanding higher temperatures and tolerant to leaf-miner. Many strains are susceptible to the die-back disease (stem-pitting). It has higher resistance to this disease than 'Bourbon'. 'Kents' has lower keeping quality. It is being cultivated in India because of judicial spraying and periodical pruning.

S.288. Released in 1938, it is the first-generation progeny of 's.26', a hybrid from Doobla estate (Western Karnataka). Its fruits are round and oblong, coated, blue to grey. Its liquoring quality is like normal arabica. 75% of the plants are resistant to races I and II of the coffee rust.

S.947. This variety (in the second generation) has been found to be homozygous for resistance to races I and II of the coffee-rust. It is preferred in Mudigere and Saklaspur zones.

S.333. It is the first filial generation (F1) of a cross between S.31 and S.22. Plants are vigorous and spreading. Fruits are bold-medium, round to oblong. About 10 % of the seedlings may show temporary chlorosis (absence of chlorophyll in the leaves) in the second or third pair of leaves. About 75 % of the plants are resistant to races I and II of the leaf rust.

S.795. Released in 1946, it is the F2 of S.288 of 'kents arabica'. Some plants are initially semi-erect but droop as they grow. Fruits are oblong with a small to broad navel flush and sometimes with persistant calyx. Beans are bold, larger than of 'S.288' and are coated or bluish grey.

S.1934. (F4 of S.288 of 'kents arabica') was established in 1960-61. They have fruits and seed characters similar to those of kents. The rest of the characters are like 'S.795'. They have a lower triage (false polyembryony). Yields are satisfactory.

Robusta. *Coffee canephora* Pierre ex Horn (2n = 22) is a bigger tree than arabica, with broader and larger leaves. Leaves are pale green, flowers are white and frogarant. They are borne in large clusters than those of arabica. Berries are small, but per node they are more numerous varying from 40-60 or more. Under the soil and climatic conditions of south India the buds appear during Nov to feb. Rains in Feb-Mar are ideal for blossoming. The flowers open on seventh or eighth day after rains. Unlike arabica, Robusta is self-sterile and hence it is cross-pollinated. The fruits mature in 10-11 months. They are harvested 2 months after arabica.

Tree Coffee. *Coffee liberica* Bull ex Hiern (2n=22) is a large tree bearing broad, dark green, leathery leaves 2-3 times larger than those of arabica.

Crosses. *Cioccie* and *Agaro* crossbred separately with 'S.1934' give good yields. Devamachy of S.881 arabica is a spontaneous *C. canephora* of *C.arabica* spotted in a private estate in Coorg. S.881 is a collection from Rume Sudan and was established at the Central Coffee Research Institute in 1944. The parent plant in Devamachy is an intermediate plant with higher resistance to rust. Beans are uniformly small. The liquoring quality is of arabica type.

S.274. robusta of 'Kents' arabica. Seeds from the selected F2 progeny of BC2 with arabica given for trial show variation from arabica to intermediate type of plants. The progeny shows an early floral phase like robusta and spreading habit like arabica. The pulp is like arabica or robusta. The bean size is variable, it is thicker as in robusta. It is mostly coated, grey, green or a mixture of the two.

Dwarf hybrids (San Ramon hybrids). established in Brasil, it has compact internodes. Seeds obtained from a cross between dwarf of s.795 of Agaro are being tried in the field. The progenies from the dwarf hybrids produce 70% dwarfs. Dwarfs of San Ramon x s.795 x Agaro, Cioccie produced dwarfs of larger size and erect branches giving an open type of plant habit. They are drought resistant and yield good crops.

Hibrido de Timor. It is a spontaneous cross of robusta x arabica from Timor supplied by the Coffee Rust research centre, Portugal. Very few plants are susceptible to races XXII,XXV,XXVI. The progenies show a high variation in morphological characters.

Planting

The soil used for raising seedlings should be rich in organic matter, free from nematodes, cockchafer larvae and wireworms. Farmyard manure, sand and lime are also important ingredients of the soil. The seed is treated evenly with Agrosan at 20 kg per forlit before sowing. This protects the soil against fungal infection during germination.

A seedbed raised to a height of 15 cm, usually about 1 metre in width and of convenient length is prepared. Four baskets of fully decomposed cattle manure or compost, about 2 kg of finely sieved lime and 400 g of rock phosphate are incorporated into a bed 1m x 6m. If heavy soils are used it is necessary to add coarse sand to facilitate drainage and aeration.

Seeds should be sown in December or January. The seeds are placed on the soil surface 1.5-2.5 cm apart in regular rows. A thin layer of fine soil is used to cover them. The bed is then covered with about 5 cm of paddy straw to ensure an even temperature and prevent the soil surface from drying. The beds are watered daily and protected from sunlight by an overhead pandal. The seeds germinate in 45 days.

Planting in the field

Disease-free and vigorous seedlings are selected for planting. Rooted plants 16-18 months old, with and without balls are planted during June and the ball and bag plants are generally planted from September to October. The seedling is placed in a hole made in the pit, with its tap root and lateral roots spread out properly. The hole is then filled with soil. The seedlings are provided with cross-stakes to prevent damage from winds.

Ball and bag seedlings are planted towards the end of the monsoon rains and with the commencement of north-eastern rains i.e. in September. First the bottom portion of the bag is cut and the tap root nipped. The seedling is gently removed from the bag and placed into the hole. The hole is covered with soil and the plant is firmly fixed.

Planting shade trees

Dadap is commonly used as a lower canopy shade tree. 2 metres long stakes are fixed for every 2 plants of coffee. Silver oak and dadap are planted during June when the rains commence. During the dry season the stems of the young dadap are either painted with a dilute solution or wrapped up in agave leaves to protect them from sun-scorch.

After care

1. The clearing should be well fenced to prevent damage to coffee and dadap from stray cattle.
2. The plants should be protected against cockchaffer attack during the first 2 years by applying Lindane or granular insecticides around the seedlings.
3. Weeds should be eradicated in the very first year.
4. The soil around the seedlings should be properly mulched and the plants provided with shade.
5. No manure needs to be applied till May-June of the following year. Sprinkler irrigation may be given.
6. It is advisable to spray a Bordeaux mixture along with urea and potash during December to protect against pests and disease.
7. A manurial schedule should be followed from the second year onwards.
8. Plant protection measure should be adopted meticulously.

Nutritional Requirements

Coffee is a perennial crop and the tree has the dual function of maturing the berries and producing fresh wood for the succeeding crop. One tonne of clean coffee removes from the soil 34 kf of N, 5 kg of P_2O_5 and 45 kg of K_2O in case of robusta varieties. Considerable nutrients are also lost through leaching under a heavy rainfall and as a result of fixation and immobilisation of nutrients in the soil. Such a depletion may lead to the impoverishment of the soil. It is thus essential to replace the lost nutrients by using fertilisers.

Shade and Shade Management

A shade tree should be fast-growing and spreading to allow a uniform amount of filtered light. Its root system should be deep so that it does not compete with the coffee plant for food and moisture. The trees, which serve as alternative hosts to the major insect pests of coffee must be avoided.

Dadap (Erythriana lithosperma) is by far the universal temporary shade-tree in India. It is always planted with coffee in the clearings. It is easily raised from stakes, has very quick growth, stands frequent loppings, regenerates very quickly and is a legume. Permanent shade trees are generally planted about 12 to 14 m apart. It is desirable to plant a large number at first and gradually thin them out. They should have their canopy about 10 to 14 m above the coffee plants. Shade trees require constant attention by pruning and lopping to furnish the required shade. Shade trees are usually raised from seeds and also from stakes. Seedlings are grown in the nursery for a year before being transplanted in the soil.

Harvesting and Processing

Quality is a summative index of many characteristics of coffee, such as its appearance in the raw, roast tastes and its liquor qualities comprising factors like aroma, body and acidity. Quality can be influenced by nutritional factors and weather conditions during the development stages of the beans. Correct processing techniques are essential to prevent deterioration in quality. Coffee is processed in two ways:

(a) wet processing by which plantation or parchment coffee is prepared, and

(b) the dry method by which cherry coffee is prepared.

Parchment coffee prepared by using the wet method is generally favoured by the market. Cherry coffee, due to its very nature of preparation and longer contact with the mucilage and fruit skin, is usually associated with a characteristic fermenting flavour.

For the preparation of both parchment and cherry types of coffee, picking of the right types of fruit forms an essential part of processing. Coffee is picked when the fruits are just ripe. Under and over-ripe fruits cause deterioration in quality, the former gives "immature beans" and the latter "foxy" coffee. They may be dried seperately for making cherry coffee. The bags used are frequently washed and dried. Bags in which fertilisers, pesticides and fungicides are stored should never be used for this purpose.

Preparation of Parchment Coffee

Pulping

The preparation of coffee with the wet method requires pulping equipment and clean water supply. Fruits are pulped on the same day to avoid fermentation. The pulper should be checked everyday to prevent cuts. The pulper-nipped beans and other deformed beans will result in defective parchment.

Fruits may be fed into the pulper through a sipho arrangement to ensure uniform feeding and to separate lights and floats from sound fruits. Uniform feeding ensures the proper removal of skins and prevents cuts, choking of the pulper etc. The pulped parchment should be sieved to eliminate any unplugged fruits and skins.

The skins separated by pulping are led away from the vats into the collection pits, so that their microbial decomposition will not affect the bean quality when they get mixed with the bean.

Demucilaging and washing

The mucilage on the parchment skin can be removed by using any one of the following methods:

(a) natural fermentation,
(b) treatment with alkali, and
(c) frictional removal in machines like 'Raoeng' and 'Aqua pulpa'.

Natural fermentation is the most commonly used method for demucilaging coffee. The mucilage breaks down in the process of fermentation. In the case of arabica, it is complete in 24 to 36 hours. Fermentation takes longer in cool water than in warm weather. If the parchment is under fermented, the sticky mucilage is left on the parchment. This condition leads to the absorption of moisture by the bean and to mustiness in the final product. When correctly fermented the mucilage comes off easily and the parchment does not stick to the hand after washing. The beans feel rough and gritty when squeezed by hand.

The robusta coffee has more of sticky mucilage. Fermentation will not be complete even after 72 hours. Quite often the mucilage breakdown is not complete even after a very long period. Thus we resort to alkali treatment or frictional removal of the mucilage. When the mucilage breakdown is complete clean water is let in and the parchment is washed pebble-clean with 3 to 4 changes of water.

Treatment with alkali

The removal of mucilage by treating with alkali takes about one hour for arabica and one and a half to two hours for robusta. The pulped beans are drained off excess water and furrowed with gorumanes (wooden ladle with long handle). A 10 % solution of caustic soda is evenly applied to the furrows using a rose can. About one kg of sodium hydroxide dissolved in 10 litres of water is sufficient to treat 25 to 30 forlits of wet parchment. The parchment is agitated thoroughly with gorumanes as to make the alkali come into contact with the parchment and is trampled for about half an hour. When the parchment is no longer slimy, and rattles, clean water is let in and the parchment washed clean with 3 or 4 changes of water.

Removal of mucilage by friction

There are pulpers such as 'Raoeng' and 'Aqua Pulpa' which pulp and demucilage the beans in one operation. A number of naked and bruised beans may occur in the parchment. It is thus necessary to adjust the machines to obtain uniform pulping and demucilaging. The sorting of fruits for uniform feeding by a siphon arrangement may also rectify this to a considerable extent. These machines are also often used for demucilaging after removing the fruit skins in the traditional pulpers.

Underwater soaking

Wherever water supply is abundant and additional vats are available, the parchment is soaked for 12 hours and then given a final wash. This improves the quality both in appearance and in the cup.

Drying

The next stage in processing consists of drying the parchment in the sun until the moisture content is sufficiently reduced to permit the storage of beans till they are despatched to curing works.

When coffee is being dried it is necessary that all naked beans, pulper-nipped and bruised beans are sorted out and despatched to curing works seperately. The coffee is then bagged into clean new gunnies. The coffee of different lots should be bagged seperately. New gunnies should always be turned inside out and well aerated before use, as otherwise coffee will absorb natural oils and off odours from the bag and give rise to an 'acrid' cup.

Cacao

The cacao plant is a perennial dicotyledonous plant which has been under cultivation since pre-historic times. The centre of origin has been placed in the tropical forests of the Amazon Valley in South America. It is now extensively grown in the continents of Africa, Asia, South America and in parts of North America.

Cacao belongs to the family Sterculiacae and has a chromosome number of 2n = 20. It is a small evergreen tree growing to a height of about 9 metres when full-grown. The plant has an unusual but characteristic mode of branching. The 2 types of branches are (1) fan or jorquette which is forked into 3, 4 or 5 and (2) 'chupon' which arises just below the jorquette and grows to increase the height of the plant.

The leaves are oblong 20-60 cm length and pointed at both ends. The young leaves are bright green or of various shades of red. The chupon type of branch bears its leaves in 3/8 spiral formation. The petiole has a cuspicuous pulvinus facilitating the turning of the leaf in response to photosynthetic stimuli. The stipules of the last-formed leaves of a flush leave scars on the tip. The leaves live through 2 flushes and are dropped from the third length back from the terminal bud so that leaves of 3 distinct ages can be seen on each branch.

Cacao is cauliflorous. Flowers and fruits arise at cushions which were originally leaf axils. The flower is pink to whitish, reguler, hermaphrodite and has 5 sepals, 5 petals, ten stamens in 2 whorls of which only one is fertile and a superior ovary of 5 united carpels. The andrecium consists of 5 long pointed staminodes and 5 fertile stamens which are positioned opposite to the petals. Flower abciss at the constriction at the base of pedicel.

The pollination mechanism in cacao is not well understood. The structure of the flower is unsuitable for pollination by natural agencies. Owing to the lack of nectar, sticky nature of the pollen which prevents wind pollination. Insects are the chief pollinating agency and a certain amount of self pollination is effected by crawling insects. Unpollinated flowers are usually shed.

The fruit is a berry containing 20 to 40 seeds each surrounded by a pulp which is a transformation of the outer integument of the ovule. The outer cells release a highly mucilaginous substance at full growth and is one reason for preventing fermentation of beans after harvesting.

Climate and Soil

Cacao is cultivated on commercial scale in the tropical lowlands where the maximum temperature does not fall below 15° celsius, the absolute min not below 10° celsius. The main production centres are confined to 10° north and south of the equator. It is grown upto altitudes of 1000 metres where the low temperatures are the limiting factor.

The rainfall should be uniform through the year a min of 90 to 100 mm per month is required and an annual precipitation of 1500 to 2000 mm unless supplemented by irrigation. The micro climate is significant. The trees prefer cool moist soil and air around them. The shade requirements are related to the soil nutrient. On a poor soil too much light may affect the yields adversely.

Cacao thrives on a wide range of soil types, with pH ranging from 4.5 to 7. It requires a deep soil. Impervious soils or those with hard substrata are unsuitable. Clayey soils or mixtures of clay and sand are preferable. Ill-drained soils are very inimical to cocao trees. Divalent bases, calcium and magnesium play a significant role in the nutrition of cacao.

In India where a monsoonal type of climate exists it is difficult to find ideal locations comparable with those in West Africa. There are a number of relatively small areas in various parts of southern India where conditions are suitable. Normal irrigated land is too expensive for a crop at present. The more suitable areas in southern India are those which get rains during the major part of the year and have a very short dry season in the regions:

- — Kanyakumari district of tamil Nadu
- — parts of Tirunelveli District of Tamil Nadu
- — Palni Hills of Tamil Nadu
- — Nynad, Trivandrum, Quilon and Kottayam in Kerela state
- — Chikmagalur District in Karnataka
- — parts of South Kanara and North Kanara Districts in Karnataka.

Cultivation

Cacao is grown in regulated shade. Development of one main trunk only is allowed. The trees are grown close so that the foliage forms a canopy to keep the air cool and damp around the trees. Regular weeding before harvesting and removal of unwanted basal 'chupons' is taken up.

Nursery

Fresh beans are to be used for sowing as they lose their viability after the pod is broken. They are planted with the pointed end upwards, sown either in bamboo baskets or polythene bags or raised beds. If sown on raised beds the young seedlings are usually transplanted in containers about 2 weeks after sowing. The seedlings will be in the nursery for 6 to 8 mths before they are transplanted to the main field. Other methods of propagation include grafting, budding and layering.

Cacao can be cultivated as an intercrop in coconut and arecanut gardens. Mixed plantations can be raised on an area 4mx 4m the cacao seedlings occupying the centre of the square or with the normal spacing of 2.7m x 2.7m for area. Pits of $90cm^2$ with compost are used for planting the seedlings. Both require artificial shade during the first one or two hot-weather periods. Banana can be profitably grown as a shade crop during the first 2 years. Subsequently, the shade cast by the arecanut trees will be sufficient for the cacao plants. The cacao plants can be manured with 100g of N, 40g of P2O5 and 140g of K2O per tree annually in 2 split doses. The plantation is irrigated at weekly intervals in the hot months. All the fan branches are cut to a height of one metre from the ground level. This operation is very important for obtaining good yields.

In coconut plantations spaced at about 7.5 metres cacao may be planted either in a single row alternating with coconut or in double rows in triangular positions between the rows of coconut. This crop combination of cacao and coconut has synergistic effect on the yield of both the crops.

Harvesting and Curing

Flowering in cacao begins in the second year of planting. The pods mature in 6-8 months. The pods when ripe are yellow to orange red. There are 2 main flushes every year i.e. in Sept-Feb and in May-Aug. The former is the main crop and the latter is the minor crop season. Mature pods contain 20-50 beans embedded in white pulp or mucilage. Ripe pods are plucked with a knife without damaging the cushions. The pods are kept overnight and opened the next day.

The beans after separation are to be fermented for removing the pulp and increase the flavour. The fermentation or sweating process differs in different localities. The wooden boxes have a perforated bottom plate to allow air for fermentation and drainage of the juice. The beans are kept covered by dry leaves to maintain the temperature required for fermentation. After 24 hours the beans are mixed and again kept in a box for fermentation. Then they are cleaned and washed and dried by spreading on a mat in the sun. The temperature for drying is 45 to 60 deg celsius. The quality and flavour of the product depends on the proper management during drying. Then the beans are fried and outer coat removed. The thick fat is removed by applying pressure on the maricated beans. The cake is powdered to prepare the cocoa powder. Theobromin is extracted from the cake.

Yield

The average yield is 600 kg of dry beans per hectare. Under mixed cropping conditions with arecanut and coconut, the plants commence flowering in one year after planting and give a high yield of about one tonne of dry beans per hectare. Cacao is both a beverage and a food crop. It is used in manufacturing chocolate, cocoa and other foods. The rind of the pods is used as cattle feed.

Varieties

The first type originating in venezuela was called 'Cariollo' and was the highest quality in the European markets and the later variety 'Forastero' or 'Trinitario' fell into the inferior quality. The Ecuador cocoa is quite different from Forastro and called 'nacional' of indigenous status.

We have 3 groups which are also equal to the market grades:

1. *Criollo cacaos.* They include Venezualan cacaos and also those which are native to Maxico or central America and Columbia. The pods are yellow when ripe, deeply furrowed, warty and pointed with thin pod wall and plump seeds. Fresh cotyledons are either white or pale violet.
2. *Amazonian Forasteros.* They are cocoas of Brazil and West Africa. Pods are yellow, less deeply furrowed, round ended with woody pod wall difficult to cut. The seeds are flattened and fresh cotyledons are violet and black. The seed characters are the most important in this case also.
3. *Trinitarios.* this is a botanically complex group which is harder but of inferior quality. The members of this group differ in parentage. For convenience they can be called hybrid mixtures of recent origin.

Rubber

Natural rubber is found in the latex of as many as 895 species of plants belonging to 311 genera of 79 families. Of these *Hevea brasiliensis* (Euphorbiacaeae), the para rubber tree is the most important source of rubber. Others are glaziovii (Euphorbiacaeae), *Ficus elastica* (Moracaeae),

Parthenium argentatum and taraxacum kok-saghyz (compositae). Many other species such as *euphorbia intisy* (Euphorbiacaeae), cryptostegia grandiflora, and landolphia sp. have been tried as possible minor sources of natural rubber.

Hevea brasiliensis is a native of Brazil and was introduced in tropical Asia in 1876 throught he Kew Gardens (England). The tree is now grown in the tropical regions of Asia, Africa, America. It is a hardy, quick growing tree with a straight trunk producing branches 3-5 m above the ground and forming a spreading canopy. The optimum ecological requirements consist of a fairly distributed rainfall not less than 200 cm, a warm humid climate (21 to 35 deg celsius) and a well-drained deep loamy soil. It flourishes from the sea level upto altitudes of 450-600 m.

Propagation

The earlier plantations were raised from unselected seeds. The yielding capacity of these trees being low the production was poor in many of the plantations. Hevea is propagated through seeds and only by adopting vegetative methods. The viability of the seeds being short, it is necessary to sow them as soon as possible. Planting is done in specially prepared nursery beds. The soil is dug up, mixed with fertilisers and beds of 30-120cm width and of convenient length are prepared. The germinated seeds are then planted about 30 x 30cm apart. The seedlings are kept in the nursery till they attain a diameter of 20 to 25 mm above the collar.

The seedlings are usually transplanted in the field in June-July. Healthy seedlings are selected and stumped at a height of 50cm. After pulling out the roots of the seedling are trimmed and then planted in position in prepared pits about 4.5m apart to accommodate an initial stand of about 500 plants per hectare. Sometimes the seeds are sown in situ in the fields, using 2 or 3 seeds per pit.

Bud grafting is done using buds from selected mother trees. Budwood nurseries to supply budwood stumps are established simultaneously with the rootstock nurseries. Nursery seedlings can be bud grafted when they attain the required girth above the ground. Budded stocks are ready to be stumped about 4 weeks after budding. They are cut 10 to 15cm above the bud patch and the stumps are pulled out and transplanted in the field. Green budding with the buds stripped from green shoots has also been practised. The plant density recommended is 420 to 450 plants per hectare. After planting the field it takes 7 years for the trees to come into tapping. The fitness for tapping is guaged by the girth of about 55cm and a height of 125 cm from the bud union. When 50-70% of the trees have reached this size, tapping is commenced.

Maintenance

The majority of the rubber trees grown in India are confined to a narrow belt from Kanyakumari district to the coorg district of Karnataka in the north and in the west of the Western Ghats. They are grown mainly in lateritic soils which are very porous, well drained and acidic. A ground cover of a leguminous crop is raised between the trees to prevent soil erosion, keep down soil temperature, conserve soil moisture and to add humus and nitrogen to the soil. The clean weeding of plantations is not now recommended.

Rubber trees should be regularly manured with balanced fertiliser mixtures from time of planting to ensure maximum production. During the first year of planting, single dose of 225g of 10:10:4:1.5 NPKMg mixture is applied to each plant. The dosage increases

900g and 1100g in two split doses during the second and third years respectively. From the fourth year onwards the application of 900g of the mixture per plant in two split doses is recommended. For trees under tapping N:P:K 10:10:10 mixture at the rate of 900g per tree is applied annually during April-May.

Abnormal leaf fall caused by *phytophthora pulmivora* and *P. meadii* the powdery mildew caused by oidium hevea and the pink disease caused by corticium are some of the major diseases affecting rubber in southern India. These disease form a limiting factor in the growth of the trees and the yield of latex, and plant protection methods have to be adopted. Prophylactic spraying with 1% Bordeaux mixture or the oil-based copper oxychloride is the control measure recommended against abnormal leaf fall. Powdery mildew is controlled by dusting with sulphur (70:30 sulphur talc mixture). The pink disease is checked by using Bordeaux paste in the early stages of infection.

Tapping

Latex is obtained from the bark of the rubber tree by tapping. Tapping is a process of wounding during which thin shavins of the bark are removed to induce flow of latex. The bark of the rubber tree has an outer hard portion and an inner zone which is soft. The latex vessels are arranged in a series of concentric rings within the bast. Their number increases towards the cambium. They are disposed almost vertically at a slight angle to the right.

Tapping cuts extends to half the circumference of the tree and leads to a vertical groove at the base of which a zinc spout is fitted. Below this a coconut shell is placed to collect the latex. Tapping commences early in the morning when the latex flow is high. After 2-3 hours when the flow stops, the flow stops the latex is collected in buckets and taken to the collection shed or factory. The tapping should be deep but the cambium is not cut. Tappings are done once in 2 days for budgrafted plants and half spiral once in three days (S/2 d/3) for clonal seedlings.

A higher intensity of tapping often leads to brown bast, physiological disorder. However on older trees, intensive tappings are practised. Synthetic hormones are also used to stimulate yield. A new yield stimulant "Ethrel" when applied almost doubles the yield of rubber. The yield of the rubber tree varies according to the type of material planted, the age of the tree, the fertility of soil, the climatic conditions and the skill of the tapper. During the south westerly monsoon many tappings are lost, however by fixing polythene rain-guards the non-tapping period can be reduced. During the wintering and refoliation period a rest of about 4 weeks is given. A rubber tree can be tapped for 25 to 30 years.

Processing

The important forms in which the crop from rubber plantations is marked are the following:

1. Preserved latex and latex concentrates
2. Dry ribbed sheet rubber
3. Dry creped rubber
4. Dry solid block rubber.

The crop collected in the form of liquid latex can be processed into any of the above forms. But the crop collected in the form of tree lace, shell scrap, and earth scrap is processed only into crepe or solid block rubbers. Preserved latex concentrates are generally in 2 concentrations— The latex between 36% to 50% drc and that between 51% and 60% drc. Latex is concentrated either by creaming or centrifuging. It is preserved by adding ammonia and other chemicals to prevent coagulation and deterioration.

The major quantity of natural rubber produced in the country is marketed in the sheet form. When the latex reaches the factory, it is weighed, seived and estimated using a metrolac (hydrometer). The latex is then standardised by adding water, allowed to stand and again seived. Then it is poured into aluminium trays of std size or into bigger tanks to which formic acid or acetic acid is added to coagulate it into a soft mass, washed in running water and passed between rollers to squeeze out the water. Finally the sheets are passed between grooved marking-rollers. The wet sheets are hung on reapers in the shade to get rid of the excess water and then moved into smoke houses for drying.

After 4 to 5 days at a temperature of 43 to 60 deg celsius the sheets are graded and packed for dispatch to manufacturers. Besides the smoked sheets pale crepe and smoked crepe are also made using different processes. In order to improve the image of natural rubber in the world market, new methods of processing and presentation are also adopted. This involves coagulation, size reduction, dewatering, dirt removal, drying baling and grading.

Varieties

Many clones have been evolved in South-east Asia and South America. The eastern clones are high yielding and named after the plantation from which they have originated. Most of the introduced clones in India are from Malaysia (RRIM), jaya(tjir, BD), Sumatra (AVROS), and Sri Lanka (mil, wagga, hil). Indian clones developed by the Rrubber Research Institute of India (RRII) are comparable with the foreign clones and are now being distributed for small-scale planting. The potential yield capacity ranges from 2000 to 3000 kg or more per hectare per year.

The production of natural rubber during 1973-74 in India amounted to 125153 tonnes and the consumption is 130302 tonnes.

Coconut

Coconut is a majestic perennial palm. It is grown in numerous islands and also in the humid coastal tracks of tropical countries. India ranks second in the world production of coconuts 6088 million nuts from 108 million hectares.

The coconut palm, or 'kalpa vriksha' provides many necessities of life like food and shelter. Of all the tropical palms it is the outstanding one. It is mainly cultivated for the nuts from which 2 important products 'copra' and fibre are obtained. Copra yields oil and oil-cake. The trunk of the mature palm is used as timber for houses and the leaves are used for thatching houses, fencing etc. The unopened spathe is tapped for toddy called 'neera'. Sweet toddy can be converted into jaggery and sugar, fermented toddy is a mild alcoholic drink and vinegar can be made from it. Water from tender coconut is a refreshing drink.

In Kerela the extraction of coir from the husk of nuts and the manufacture of coir products provides employment for thousands of people. The coconut shell is largely used for fuel and for the production of charcoal and making a variety of curios. The shell flour is used as a filler for plastics. Thus every part of the palm is useful in one way or the other.

Climate

Coconut is essentially a crop of the humid tropics. It is mainly grown in the coastal plains. Rainfall is the most important factor affecting its growth. A rainfall for 100 to 225 cm per annum, evenly distributed is most important.

Coconut requires an equable climate. The optimum mean annual temperature for its best growth and maximum yield is 27 deg celsius. Frost and low humidity adversely affect the yield of the palm. Persistent high humidity is also harmful and bud rot is more common under such conditions. The palm requires bright sunshine of 2000 hrs a year.

Soil

The coconut palm adapts itself remarkably well and thrives in almost all types of well-drained tropical soils, such as coastal sand, red loam, laterite, alluvial and reclaimed soil of the marshy lowland. A water table that is too high and remains static for too long is harmful.

Planting Material

Since coconut is a cross pollinated perennial crop the selection of seeds is of vital importance. Selection has to be made at the mother-palm level and at the seedling stage. The mother palms should be healthy, high yielding and regular in bearing. Also factors such as proportion and distribution of female flowers, high setting percentage and a high copra content should also be taken into consideration. If prepotent palms can be identified by progeny testing the yielding ability of the planting material can be further assured.

A careful selection of the seedlings is important. The seedlings should be healthy and have a minimum of six leaves when they are one year old. The early splitting of leaves is a desirable character. Nine month old seedlings having a minimum of four leaves can also be planted.

Preparation of Land

The clearing and levelling of land have to be done before preparing the pits for transplanting seedlings. The depth of the pit depends on the soil type. In the sandy loam of low water-table the planting pits of the size 1m x 1m x 1m is recommended. In laterite soils with a rocky substratum, deeper and wider pits 1.2 x1.2 x1.2m are necessary. At the time of planting the pits have to be filled up with loose soil leaving a final depth of about 60cm. However when the water table is high planting on the surface or on soil mounds may be necessary. In a littoral sandy soil, circular pits 1.5m x1.5m x1.5m are made and partially filled with alternate layers of red earth and coconut husks (150 husks).

The spacing of palms requires careful consideration. A spacing of 7.5 to 9 m accommodating 177 to 124 palms per hectare is adopted under the square system of planting. The triangular method accommodates 20 to 25 more palms per hectare. In the hedge system of planting, a spacing of 5 to 5.5m along the row is given. Larger number of palms can be planted in a unit area ensuring sufficient light on each palm.

Transplanting and After Care

In well drained soils seedlings can be transplanted at the beginning of the south-westerly monsoon. If facilities for irrigation are available, it is advisable to take up planting at least a month before the onset of the monsoon. In the low lying areas, seedlings are better transplanted after the rainy season. The transplanted seedlings should be shaded and irrigated properly during the summer. Irrigation with 45 litres of water once in four days is the optimum in sandy soils. Provision for drainage is also important. Apply the complete NPK fertiliser from the early stage. The first dose should be applied 3 months after planting.

Large quantities of green leaf or compost to the soil can be applied where it has a poor organic matter content. The pits should be cleared of all the weeds and the soil washed down by rain. The palms should be frequently examined for insect attack or fungus diseases and the necessary remedial action adopted promptly.

Care of Adult Palms

The palms generally start bearing at the age of five or seven years after planting and the stabilised yield is obtained from about the tenth year onwards for 50 to 60 years. Regular intercultivation and manuring of the palm are essential for stepping up and maintaining the productivity of palms at a high level. Tillage including digging, ploughing, piling of mounds, levelling of mounds, etc. and the making of shallow basins each with a radius of 2 m, at the beginning of the monsoon and filling them up at the close of the monsoon are beneficial. In sandy soils, which are of low fertility and do not have a luxuriant growth of weeds, regular intercultivation may not be quite necessary but in other types of soils, intercultivation is essential for keeping down the weeds under check and for creating a soil mulch.

The method of intercultivation mainly depends on the local conditions, soil type, topography, size of holding, intencity of rainfall, etc. The recommended dose of fertilisers is 500g of N, 300g of P_2O_5 and 1,200g of K_2O per palm per year for the ordinary tall variety. For soil which are poor in organic matter, the application of green manure or compost at 50kg per palm is recommended. Green-manuring crops usually grown are cowpea, sunnhemp. Fertilisers may be applied in two spilt doses. After the receipt of summer showers one third of ht edose of fertilisers is spread around the palm within a radius of 1.8m and raked in.

Circular basins 1.8m in radius and 25cm in depth may be dug in July-August and the green leaf manure or compost applied and partially covered. The remaining two-third dose of the fertilisers may be spread in the first week of September and the basins closed. Under irrigated conditions and in areas with assured supply of moisture throughout the year the annual dose of fertilisers may be applied every three months in equal split doses.

In regions with a long spell of dry weather, there is a definite set back in the growth and yield of the coconut palm. An increase of 50% to 100% in the yield of nuts can be obtained by giving summer irrigation. The quantity and interval between irrigations depend upon the soil and the climatic conditions. In loamy soils copious irrigation, once a week is sufficient. In sandy soils along the sea coast during summer without any adverse effects

Mango, jackfruit, arecanut and other crops and a number of annual crops are commonly grown along the coconut palms in household gardens. Recent studies have shown that under rainfed conditions tuber crops, banana, pineapple, upland rice and pulses can be grown successfully as intercrops in adult coconut plantations without any detrimental effects on the fertility of the soil and the productivity of the palms. Under irrigated conditions cacao forms a very beneficial crop combination with coconut.

Other crops suitable for mixed cropping include pepper and tree spices, such as cinnamon, clove, nutmeg. Mixed farming including the raising of milch animals on fodder grass and legumes raised as intercrops are manured with cowdung and urine. This practice improves the soil fertility by recycling organic wastes, enhances the yield of palms and generates additional income and employment potential for the farmer's family.

Yields

The coconut palm is unique in that once it attains the normal bearing stage, it continues to bear a bunch of nuts in every leaf axil almost at monthly intervals all year round and throughout its life extended over 50 to 60 years. The nuts mature nearly one year after fertilisation. Generally, harvesting is done once in 45 to 60 days. Tender nuts in demand as a delicious drink in West Bengal and Maharashtra are best harvested at the age of 6-7 months. Large scale harvesting of the tender nuts is not being done in other parts of the country. Nuts for culinary purposes, for making cup copra and good quality fibre are harvested at the age of about 11 months whereas those for ball copra and coarse coir are harvested only when they are fully ripe.

The average yield per hectare varies from 10,000 to 14,000 nuts per annum. From a well maintained garden an annual yield of 25,000 nuts per hectare can be obtained. Nearly one-third of the annual yield is harvested during the three months— march, april and may—in conditions obtaining on the west coast.

Coconut Products

The main product obtained from the coconut is copra or the dried kernel of the nut. The cup copra is prepared by splitting the husked nut into two halves and drying them in the sun. After drying for a day or two the kernel gets detached from the shell. The drying is continued for another four to five days. Ball copra is prepared by storing fully ripe nuts for 8-12 months on a raised platform in a shed. They are sometime smoked or heated with a slow fire to accelerate drying. During storage water inside the nuts gets dried up and the kernel gets detached. Nuts are then husked and the shells are carefully broken. The ball copra thus obtained is clean, sweet and highly valued.

The coconut fibre is known as coir. For retting, husks are buried in shallow pits in low-lying areas subject to the tidal flow of backwaters for about 6-12 months. The coir obtained from dry husks would be of inferior quality. It is easier to extract coir from green husks after sufficient retting and the quality of coir is good. The coir is then dried in the sun and spun into coir yarn by hand or with a machine. Coir yarn is used as such or made into ropes, mats, nets, bags etc.

Varieties

There are two distinct varieties of coconut the tall and the dwarf. Dwarf x tall and tall x dwarf hybrids have been found to be early bearers and superior to their parents in yield potential under favourable conditions. The tolerance of these hybrids to the root wilt disease adds to their virtue.

Arecanut

Arecanut or betelnut is an extensively cultivated tropical palm the nuts of which form a popular masticatory in India, the Middle east, and Far East. India produces annually 150,000 tonnes of arecanut from an area of 18,34,000 ha. It is a tall stemmed erect palm, reaching varied heights, depending upon the environmental conditions. Palms attaining a height of 30 metres are not uncommon.

Climate

The altitude at which the arecanut palm can, be successfully grown varies to some extent according, to the latitude of the place. Though it grows at altitudes up to 1,000 m above the sea-level, at higher altitudes it is not at all productive. The cultivation is mostly confined to 28' North and South of the Equator. It is unable to withstand extremes of temperatures and wide diurnal variations. The range of temperature at which it can flourish is from about 15° to 38° C.

Areca thrives well in a variety of soils, the laterite soils of the West Coast, the red loamy soils of the Mettupalayam (Tamil Nadu), the alluvial of Assam and West Bengal and the loans of Orissa. The foremost factor that has to be considered in establishing an arecanut plantation is that the site selected should have adequate facilities for irrigation. The soil also should be deep and,well drained, without a high water-table. Being, highly susceptible to a sun-scorch, the areca palms need adequate protection from exposure to the south-western sun.

Selection and Raising of Planting-Material

It is important to select genetically superior planting-material for which mother-palms possessing characters of high heritability, correlated with high yield are to be identified. Of the many mother-palm characters studied, the age at first bearing and the percentage of the nuts set were found to be correlated with yield and having high heritability. The selection of seed nuts may be commenced only after the stabilisation of yield of the palm. This stabilisation generally takes 4 to 5, years from the first bearing. Only fully ripe nuts should be selected as seed nuts, discarding underdeveloped and malformed ones.

Nursery

The selected seed nuts are sown soon after harvest, 5 to 6 cm apart, in beds of sand under partial shade, with their stalk-ends pointing upwards. Sand is spread over the nuts just to cover them. The nuts are irrigated daily. Germination starts about 40 days after sowing and the sprouts are ready for transplanting when they are about three months old. Nursery-beds of 150 cm width and of convenient length are prepared for transplanting sprouts. The sprouts are to be transplanted at a spacing of 30 x 30cm with the onset of the monsoon. A basal dose of well-decomposed cattle manure at the rate of 5 tonnes per hectare may be applied. Partial shade can also be provided with an artificial pandal or by raising crops alike coccinia indica. The seedlings are transplanted when they are 12 to 18 months old. Seedlings having max number of leaves and min height are selected for transplanting.

Preparation of Land

The land is prepared well by digging or repeated ploughing and is levelled and terraced, if necessary. Channels are to be provided if the palms are to be grown under irrigation. The drainage of soil is to be attended to.

Pits of 90 cu cm dug at 2.7m apart both ways are used for planting seeds. The seedling is to be planted in the centre of the pit covered with soil upto the collar. The soil is to be pressed on all sides. Planting is done usually in May-June in well-drained soils, but in clayey soils, subject to water-logging, planting can be postponed till August-September to obtain a better establishing of plants. A shade crop like banana can be raised which gives good return. A proper alignment of the plantation will prevent scorching of the stem. In the square system of planting at a spacing of 2.7 x 2.7m the north-south line should be deflected at an angle 35° to the west. The outermost row of plants on the southern and south-western side can be protected by growing tall and quick growing shade trees.

Manuring and Interculture

A steady and high yield will depend much on the adequate availability of plant nutrients in the soil. This is all the more important in the case of such a perennial crop. Since almost all the arecanut-growing areas are in heavy-rainfall tracts the soils are liable to leaching and erosion thus making them poor in major plant nutrient and organic matter.

The annual application of 100g of N, 40g of P_2O_5 and 140kg of K_2O in the form of fertilisers and 12 kg each of green manure and compost and cattle manure per bearing palm is recommended. The fertilisers may be applied in two split doses.For young palms a full dose of green manure and compost or cattle-dung may be applied from the first year of planting. The second dose of fertiliser can be applied to the base of each palm all around and mixed by a light forking. The soils are mostly acidic. The application of lime corrects this, increases availability of soil nutrients, reduces the phosphorus fixation and enhances the general condition of the soil. The application of lime has to be completed at least three weeks before manuring in September-October.

The palms may be irrigated once in 3-5 days. In souhtern Kerela where arecanut is mainly grown it has been found that manuring along with irrigation gives three times more yield than manuring alone. Adequate drainage should also be provided since the palms cannot stand water-logging. The main cultural operations are performed close to monsoon generally in October-November. Where the land is slopy terracing has to he done to prevent soil erosion.

The raising of green manure crops, sucn as Mimosa invisa, Stylosanthes gracillis and Calopogonum mucunoids was also found to be advantageous. Owing to the long pre-bearing age of this palm, practically no income is obtained during the first several years. Intercropping with suitable crops, such as elephant foot-yam, banana, guinea-grass or mixed cropping with cacao, pepper and betelyine can be taken up in areca gardens without any detriment to the yield of the main crop. The culling out of uneconomic trees and replacing them with good seedlings is important in maintaining a high level of productivity (of the garden).

Harvesting

The pre-bearing age of the palm ranges from 5 to 8 years. Though inflorescence initiation has been observed in every leaf axis, there is absorption of infloroscences to about 5 per cent. The plant is monoecious, producing both male and female flowers on the same tree. The spedix of a grown-up palm produces on an average, 294 female flowers. The colour of the fruit during its growth changes from green to different shades of Yellow and red during ripening. In some places, tender cuts are harvested, whereas in others, both immature and mature nuts are harvested. Tender nuts are harvested from July to December and ripe nuts from December to March or from May to July. Three or four pluckings are, done during the whole season.

Plant Products

Arecanut is consumed both as a raw/ripe nut (adaka or Kacha tamul) as dried ripe nut (chali supari) and as semi-mature cut and processed varieties 'Batoldike" or 'Kalipak'. There are over 150 trade types, differing in maturity, processing conditions and varying in their taste characteristics as per market conditions prevailing at different centres of the country. The drying of the whole fruits for making chali supari requires up to 40 to 45 days of good sun- shine, so as to get a moisture level of about 10 per cent.

Drying ripe nuts on cement floors reduce fungal infection of the nuts to a minimum level of about 5 per cent. A drier designed recently has been found to be most suitable to produce good quality chali supari. The cup shaped nuts are prepared by boiling tender arecanuts after husking and cutting into halves. Frequant additions of the decoction commonly known as chogaru obtained by the pressure boiling of tender nuts are added to the nuts. Later the nuts are dried over mats in the sun for about 7 days. An important by-product is the husk of the nuts which can be utilised for making boards, paper etc. The spathe covering the infloroscence and the leaf sheath can be used for making caps and for packing. The palm trunk is a useful building material.

Varieties

Of the several varieties of the genus areca, A. catechu is the most commonly cultivated species. Five exotic introductions viz. VTL2,11,12,13,17 and an indigenous one 'mohitnagar' have been found to be superior to the local variety in economic attributes. Of these VTL3 has been released by the Central Variety Release Committee under the name 'mangala'. 'Mangala' is an introduction from the orient. It is a semi-tall early-bearing variety, about 70 percent more than the local variety.

6

Production Management of Oilseed Crops

Oilseed crops have been the backbone of agricultural economy of India from time immemorial. Today these crops are cultivated on about 16.5 million hectares, with total production of 10 million tonnes. This area constitutes approximately one-tenth of the total cultivated area in India. On the oilseed map of the world, India occupies a prominent position, both in regard to acreage and production. The important oilseed crops grown in this country in order of importance are groundnut, rapseed and mustard, sesame, linseed, safflower, castor, sunflower and niger.

Groundnut

Groundnut (*Arachis hypogaea L.*) is believed to be the native of Brazil to Peru, Argentina and Ghana, from where it was introduced into Jamaica, Cuba and other West Indies islands. The plant was introduced by Portuguese into Africa from where it was introduced into North America. It was introduced into India during the first half of the sixteenth century from one of the Pacific islands of China, where it was introduced earlier from either central America or South America.

Economic Importance

The oil content of the seed varies from 44 to 50 per cent, depending on the varieties and agronomic conditions. Groundnut oil is an edible oil. It finds extensive use as a cooking medium both as refined oil and Vanaspati Ghee. It is also used in soap making, and manufacturing cosmetics and lubricants, olein stearin and their salts. Kernels are also eaten raw, roasted or sweetened. They are rich in protein and vitamins A, B and some members of B2 group. Their calorific value is 349 per 100 grammes. The H.P.S. type of groundnut kernels are exported to foreign countries. The residual oilcake contains 7 to 8 per cent of N, 1.5 per cent of P_2O_5 and 1.2 per cent of K_2O and is used as a fertiliser. It is an important protein supplement in cattle and poultry rations. It is also consumed as confectionary product. The cake can be used for manufacturing artificial fibre. The haulms (plant stalks) are fed (green, dried or silaged) to livestock. Groundnut shell is used as fuel for manufacturing coarse boards, cork substitutes etc. Groundnut is also of value as rotation crop. Being a legume with root nodules, it can synthesise atmospheric nitrogen and therefore improve soil fertility.

Botanical Description

Groundnut (*Archis hypogaea L.*) is a member of sub-family, Papilionaceae of the family Leguminosae. *Archis hypogaea L.* consists of two subspecies each containing two botanical varieties.

Plants of the botanical variety hypogaea are spreading (runner) to upright (erect bunch) in growth habit, have alternate branching, lack inflorescences on the main stem, possess appreciable fresh seed dormancy, flowers are longer and mature later than those of subspecies fasligiata. Variety hirsuta has been used only to a little extent. Plants of this subspecies fasligiata are upright, have sequential branching and inflorescences in the main-stem leaf axils, possess little fresh seed dormancy, and are of shorter duration than those of the subspecies hypogaea. Subspecies fasligiata includes both the Spanish and Valencia types

Groundnut, in general, has a short-statured plant, with the main axis being upright (15 to 40 cm long) but the major part of the plant consists of the primary branches. Secondary and tertiary branches are found in the semi-spreading and spreading types, giving them a prostrate stature. The leaves are alternate, stipulate and quadri-foliate. The flowers are orange yellow, typically papilionaceous, with a long calyx tube within which is held the style borne on a superior ovary. The calyx and corolla lobes are borne in the axils of the leaves on the fruiting branches. They are bisexual, Zygomorphic, complete and sessile. Petals are five, with one large standard, two wings and two fused keel petals. There are two steering and eight fertile anthers, four of which are globose and four oblonge.

Groundnut is predominantly a self-pollinated crop and pollination takes place early in the morning. As soon as the fertilisation is complete, the flowers fade. After fertilisation, an intercalary meristem becomes active at the base of the ovary above the point of attachment of the hypanthium, producing new tissue below itself and resulting in an elongated stalk in the peg. The pegs are positively geotropic, enter the soil and bend in a horizontal plane. Generally two, and occasionally one, three or four, fertilised ovules are borne at the tip of the peg which later swells to become the pod. The testa is generally pink, but varieties with red, white, purple and blotched testa, with various gradations of colours are available.

Distribution, Area and Production

The major groundnut-producing countries of the world are India, China, Nigeria, Senegal, Sudan, Burma and the USA. Out of the total area of 18.9 million hectares and the total production of 17.8 million tonnes in the world, these countries account for 69% of the area and 70% of the production. India occupies the position, both in regard to the area and the production, in the world. About 7.5 million hectares is put under it annually and the production is about 6 million tonnes. 70% of the area and 75% of the production are concentrated in the four states of Gujarat, Andhra Pradesh, Tamil Nadu and Karnataka. Andhra Pradesh, Tamil Nadu, Karnataka and Orissa have irrigated area forms about 6% of the total groundnut area in India.

Climate and Soil

Groundnut is grown throughout the tropics and its cultivation is extended to the subtropical countries lying between 45 degrees N and 35 degrees S and upto an altitude of 1000 metres. The crop can be grown successfully in places receiving a minimum rainfall of 1,250 mm. The rainfall should be well distributed well during the flowering and pegging of the crop. The total amount required for pre-sowing operations is 100 mm; for sowing, it is 150 mm and for flowering and pod development an evenly distributed rainfall of 400-500 mm is required. The groundnut crop, however, cannot stand frost, long and severe draught or water stagnation.

Groundnut is grown on wide variety of soil types. However, the crop does best on sandy loam and loamy soils and in the black soils with good drainage. Heavy and stiff clays are unsuitable for groundnut cultivation as the pod development is hampered in these soils.

Rotation and Mixed Cropping

Generally, as a kharif crop, groundnut is grown year after year. In certain places, it is rotated with wheat, jowar, bajra, garden crops, such as potatoes, onions, chillies, garlic, ginger and turmeric. The yields of the cereal crops following groundnut are usually increased by about 25%. Pulses, e.g. red gram, mash and moong are grown mixed with groundnut. In certain places, millets e.g. jowar and bajra, and castor are grown mixed with groundnut.

Groundnut is raised mostly as a rainfed kharif crop, being sown from May to June, Depending on the monsoon rains. In some areas or where the mansoon is delayed, it is sown as late as August or early September. As an irrigated crop it is grown to limited between January and March and between May and July.

Cultivation

For a kharif crop, with one set of rains in May-June, the field is given two ploughings and the soil is pulvarised well to obtain a good tilth. The third ploughing may be given just before sowing. Harrows or tillers can be used for cultivation. When the soils are heavily infested with perennial weeds, e.g. Cynodon or Cyperus, very deep ploughing is needed. If a field is infested with white grubs, chemicals, such as Heptachl or chlordane, are drilled at the rate of 25 kg per ha before final borrowing. For the irrigated crop, beds of convenient size may be made, depending on the tropography of the land, the nature of the irrigation source and the mode of lifting water. Since groundnut is a deep rooted plant it uses up both moisture and nutrients in the deeper layers of the soil. Generally, the following fertiliser schedules are recommended, depending upon the soil fertility, the variety grown, and the quantum and distribution of rainfall.

Usually, the application of only N and P is resorted to, since most of the Indian soils are rich in K. The application of N in two equal split doses, one before sowing and the other 30 days after sowing, is found to increase the yield. In virgin lands, when groundnut is newly introduced, the application of a culture of Rhizobium as seed treatment is beneficial in increasing nodulation and nitrogen fixation. The application of gypsom at 500 kg per ha at the pegging stage will enhance pod formation.

Healthy and well developed pods should be hand shelledor shelled with a suitable groundnut sheeer about fornight before sowing. Well defined kernels should be selected and treated with 5 gm of thiram or 3 g of Captanper kg of kernels. The kharif crop is sown with a speed-drill or with a suitable planter at a depth of 8 to 10 cm. specings adopted differ from place to place. For the rainfed bunch groundnut the inter-raw specings vary between the adjacent rows varies from 20 and 30 and intra-raw specings between 10 and 20 cm. The quantity of well developed seeds required per hectare with the above specings will be about 110 kg for semi-spreading and spreading varieties and 120 kg for the bunch varieties.

Weeds cause considerable reduction in yield. A reduction of 20 to 45% in yield due to weeds has been recorded. For controlling weeds, and also to keep the soil in a friable condition, the crop should generally receive a hand-weeding, with bulllock-drawn implements, the first about three weeks after sowing and the second and the third about a fortnight and a month later. No interculture would be done after the pegs have commenced going underground. Earthing up can be done in the case of bunch and semi-spreading types to facilitate the maximum penetration of the pegs into the soil. Weeding can also be done quickly and efficiently with a wheel hand-hoe, if the soil is light. It should be run before weeds make excessive growth. Weeds can also be controlled effectively with Lasso or Tok-E-25 weedicides at the rate of 5 litres in 500 litres of water per hectare as a pre-emergence soil spray within two days of sowing groundnut. If the Kharif crop is cought in a long spell of draught, especially at the pod formation stage, supplemental irrigation is given, if water is available. For the irrigated groundnut, the frequency of irrigation depends on soil texture and interval between irrigation ranges fro 8 to 12 days. Studies on the water requirement of indicate that the period between the fifty first and fifty seventh day the crop depends on timely irrigation to it during this period.

The prominent symptoms of maturity are the yellowing of leaves, the shedding of older leaves, the development of proper colour of the testa and the dark tint inside the shell. The bunch and semi-spreading varieties are usually harvested by hand pulling when there is adequate moisture in a soil. The spreading types, on the other hand, are harvested by digging with a mammity or by ploughing or working a blade harrow. The pulled out plants are stacked for a few days for drying and stripped afterwards.

The stripped pods are cleaned and dried to a safe moisture content of not more than 5%. Damp nuts, if stored, will ferment and allow the development of poisonous moulds, e.g. aspergillus flavus in the Kernels, leading to contamination with aflatoxin—a health hazard both human beings and livestock. The oil express from such produced will be rancid and the cake. When fed to poultry results in a heavy mortality of the birds. It is desirable to store groundnut in gunnt bags are stacked in a storeroom in tires comprising not more in each in such a way that the air keeps circulating planks to avoid damage from dampness, rats, etc. The store rooms should be periodically inspected to ensure that there is no storage pest or to decompose of the stock immediately after proper cleaning.

Pests and Diseases

Aphids (*Aphis craccivora*), Leaf miner (*Stomopteryx subseciva* Zell)., the red hairy cater piller (*Amsacta albistriga* Walk., and *A moorei* Bltr.) and the white grub (*Lachnosterna consanguiea*) are the serious pests of the groundnut crop. Dusting B.H.C 10% on the young larve or spraying Metasystox) 0.05% on grown up catrepillers is recomended for controlling the red hairy caterpiller. Spot-treatment with Phorate or Mephospholan granules at the rate of 1.5 kg a.i. per ha into the soil before planting is recomended for controlling the white grub.

The tikka leaf-spot (*Cercospora arachidicola* and *C. personata*) and the coller- rot (*Aspergillus niger* and *A. pulverulentum*) are important diseases of groundnut. Rust (*Puccinia arachidis*) has been occuring in a serious form in recent years in certain groundnut growing areas. It is desirable to store groundnut in gunnt bags are stacked in a storeroom in tires comprising not more in each in such a way that the air keeps circulating planks. Dry seed dressing with Thiram or captan @ 4g/kg of seed is recomended for controlling collar-rot. Yet spraying with Hinosan at the rate of 0.02% is recommended to check the spread of the Pathogen.

Yield

Under rainfed conditions, the average yield of semi-spreading and spreading varieties is 1200-1400 kg of unshelled pods per hectare and that of bunch types is 800-1,000 kg. The crops grown with supplemental irrigations produce 3,000 kg more of pods per hectare.

The pods yield 70 to 75 per cent of kernels by weight. The yield of haulms is usually two to two-and-half times that of pods.

Varieties

The varieties under cultivation fall into three groups in respect of the habit of growth, namely bunch (Spanish), semi-spreading (Virginia bunch) and spreading (Virginia runner). In the bunch group, the plants grow erect, possess light-green foliage, produce pods in clusters at the base of the plant and have round, plump non-dormant seeds, with light-rose testa. In the case of the semi-spreading and spreading varieties, the branches trail either partially or completely on the surface of the soil, produce pods all along them, possess dark-green foliage and have oblong, dormant brownish seeds. The semi-spreading and spreading types are usually heavier yielding and later-maturing than the bunch varieties.

Rapeseed and Mustard

Under the names rapeseed and mustard, several oilseeds belonging to the crucieferae are grown in India.

1. Brown mustard, commonly called rai (raya or laha) —*Brassica juncea* (L.) Czern. & Coss
2. Sarson
3. Toria (lahi or Maghi Labi) —*B. campestris* L. var. toria Duth.
4. Taramira or tara (*Eruca sativa* Mill.)

In trade, sarson, toria and Taramira are known as rapeseed, and rai as mustard.

Banarsi rai (*B. nigra* Koch.) which does not fall under any of the four groups is a garden crop used as spice. The cultivation of white mustard (*Brassica alba* syn. *Sinapis alba*) is no longer found in India. Rai and Yellow sarson are self-fertile and the rest of the cruciferous oilseeds, viz. brown sarson, toria, taramira, Banarasi rai and white mustard, are self-incompatible. The acreage under yellow sarson is scantily (mainly in Bihar) and constantly on the decrease. In the recent past, the accreage under brown mustard is steadily on the increase (over 65%) at the expense of other Brassica due to its higher production, greater to pests and diseases and moisture stress.

Brassica juncea L. (rai) was originally introduced from China into north-eastern India. From where it has extended into Afghanistan via the Punjab. Eastern Afghanistan, together with the adjoining north-western India is one of the independent centres of brown sarson (*Brassica campestris* var. brown sarson). Yellow sarson (*B. campestris* var. yellow sarson) is commonly grown in the eastern parts of India where it shows much diversity of forms. Taramira is a relatively recent introduction into India. It is believed to be a native of southern Europe and North Africa.

Economic Importance

Rapeseed and mustard yield the most important edible oil content of the seeds of different ranges from 30 to 48 per cent. In the case of white mustard, the oil content ranges from 25 to 33 per cent. The oil obtained is the main cooking medium in Northern India and can not be replaced by any other edible oil. The seed and oil are used as a condiment in the preparation of pickles and for flavouring curries and vegetables. The oilcake is mostly used as cattle feed. The leaves of young plants are used as a green vegetable. The use of mustard oil for industrial purposes is rather limited on account of its high cost.

Botanical Description

Rapseeed and mustard include annual herbs ranging in height from 0.45 to 1.75 m. Roots, in general, are long and tapering. Toria is more or less a surface-feeder and brown sarson has long roots, with a limited lateral spread, enabling its successful cultivation under drier conditions. Yellow sarson has both extensive and lateral spread. The height of the stem varies from 45 cm to 1.75 m. In toria and brown sarson, the branches arise at an angle of 30 0 to 400. In yellow sarson, branches arise laterally at an angle about 10 0 to 200 and give the plant a narrow and pyramidal shape. The inflorescence is coroymbose raceme. In the case of sarson the four petals are spread apart, whereas in brown sarson and toria the petals overlap or may be placed apart, depending upon the variety. The flowers bear a hypogynous syncarpous ovary. In brown sarson and toria, the ovary is bicarpellary, whereas in the case of yellow sarson, it may also be tri or tetra-carpellary.

The fruit is a siliqua. The pods are two-valved, three-valved or four-valved, depending upon the number of carpels in the ovary. The flowers begin to open from 8 a.m. and continue up to 12 noon.

Distribution, Area and Production

The crop is grown both in subtropical and tropical countries. In Asia, it is chiefly grown in China, India and Pakistan. It is also grown in Europe, Canada and the USSR, but the forms of rapeseed and mustard grown there are different from those grown in India.

India occupies the first position, both with regard to acreage and production of rapeseed and mustard in the world. In India the Braddica crops occupy the second largest position after groundnut, with 3-5 million hectares, producing about 2 million tonnes of seed annually. The chief states producing them are Uttar Pradesh, Punjab, Haryana, Assam, Bihar, Madhya Pradesh, Rajasthan, West Bengal and Orissa.

Climate and Soil

The rapeseed and mustard crops are of the tropical as well as of the temperate zones and require relatively cool temperatures for satisfactory growth. In India, they are grown in the rabi season from September-October to February-March. The rapeseed and mustard crops grow well in areas having 25 to 40 cm of rainfall. Sarson and toria are preferred in low-rainfall areas, whereas raya and toria are grown in medium and high-rainfall areas respectively.

The rapesed and mustard thrive best in light to heavy loams. Raya may be grown on all types of soil, but toria does best in loam to heavy loam. Sarson is suited to light-loam soil and taramira is mostly grown on very light soils.

Cultivation

A fine seed-bed is required to ensure good germination. In irrigated areas, the first ploughing is done with a medium sized soil-turning plough, followed by two to four ploughing with a desi plough or a cultivator. Sohaga (Planking) is given after every ploughing.

In rainfed areas, one to two ploughings with a desi plough or a cultivator, each followed by planking, may be given. Toria, in particular, requires a fairly moist seed-bed for good germination, but excessive moisture should be avoided. The rate in the case of mixed cropping depends on the proportion of the rapeseed to the main crop.

When sown pure, 5 kg of seed per ha is used for all rapeseeds and mustard. When sown mixed with other crops, the time of sowing rapeseed and mustard is governed by the sowing of the main crop. The first half of September is best for sowing toria, 25th September to 15th October for sarson, 30th September to 15th October for raya, and tarantira is sown throughout October. The seed from healthy and desirable plants, grown in isolation in the case of self-sterile form, should be used. Whenever moisture in the field is inadequate, the seed is mixed with moist soil and kept overnight. For distributing evenly, the seed is usually mixed with sand before sowing.

In mixed cropping rapeseed and mustard are sown in rows 1.8 to 2.4 metres apart across the main crop. The pure crop of rapeseed or mustard is sown at a depth of 4 to 5 cm in lines, 30 cm apart, with a drill, or with a (tube) attached to the plough. Thinning is done three weeks after sowing to maintain a plant-to-plant distance of 10 to 15 cm.

One hoeing in the case of toria in the third week after sowing and 1-2 hoeings in the case of sarson and raya are adequate. Forty kg of W per ha is optimum for all rapeseed and mustard crops in rainfed areas. Under irrigated conditions doses of 40 and 80 kg of W per ha are considered optimum for toria, sarson and rapeseed respectively. For tarantira, 20 kg of W per ha is ancient. All the fertilisers should be drilled before sowing. Among the Brassicae, raya is most responsive to irrigation, followed by yellow sarson. Two irrigations, one at flowering and the other at pod formation, result in the maximum yield in the case of toria, sarson and raya. Its flowering is most economical, if there is water scarcity. Tarantira is not irrigated.

Harvesting is done as soon as the crop begins to turn yellow. Tma, which takes 75 to 90 days to mature, is the earliest crop to be harvested. Harvesting is done with hand-sickler. Threshing is done by beating with a wooden stick the seed bearing part of the plants, taken in convenient bundles or by trampling them under the feet. Winnowing is done with by slowly dropping the threshed produce from a basket held shoulder-high. The seed after being dried in the sun is stored in gunny bags or bins.

Pests and Diseases

The most serious pests of rapeseed and mustard are the mustard aphid (*Lipaphis erysinbi* (Holt) syn. *Rhopalosiphun pseudobrwsicae* (Davis), mustard sawfly (*Athalia proximo King*) and cutworm (*Agrotis ypsi"* Poott.). To control aphidy the spraying of Methyl Demeton 0.02% or 1imetboate 0.03% or Phosphomidon 0.0V/ two or three times, depending on the intensity is recommended. The mustard sawfly is controlled by dusting 10% BHC at the rate of 25 kg per ha. Cutworms are controlled by drilling into the soil 5% dust of Aldrin or Heptachlor or Chlordane at the rate of 25 kg per ha.

Alternaria blight (*Alternaria brassicae* Bork.) is the most widespread and destructive disease of rapeseed and mustard, whereas rust (*Cystopus candidus*) is serious in certain areas. Spraying with Dithane M-45 or Difolaton at the rate of 11 kg per ha is recommended for controlling these diseases. Orobanche, a phanerogamic parasite, has also been reported to be serious in Punjab and in certain parts of Uttar Flradesh. Hand-pulling of the plants of the parasite and a long rotation are recommended. Yield. The average yield of rapeseed and mustard is about 500 kg per ha. Under good crop management, toria, which is a short-duration crop, gives a yield of about 450 to 650 kg per ha, sarson 800 to 1,000 kg and rai 1,000 to 1,200 kg per hectare.

Oil and Protein Quality

The oil of rapeseede and mustard possesses a Sizable amount of crucie acid (38 to 57%), together with linolenic acid up to 4.7 to 13.0 %. The oleic and linolenic acid, which have nutritive value, together constitute only about 27 per cent. It is desirable to increase the quantity of oleic and linolic acid and erucic acid. A lower proportion of erucic acid will make the oil more palatable, nutritive besides reducing metabolic disorders.

The Protein contained in rapeseed and mustard normally ranges between 24 and 30%. On the whole-seed basis and between 35 and 40% on the meal basis. But the presence of toxic glucosinolates in the mustard cake is unsuitable as source of human protein and is at present use as a manual and as a cattle feed.

Varieties

The varieties of different forms of rapeseed differ greatly in branching habit, maturity, shape and colour of foliage, colour and size of flower, seed, etc.

Hints for Enhancing the Yields

1. Select the right crop according to the soil type.
2. Sow the seed of an approved variety only.
3. Prepare a fine seed-bed.
4. Sow the crop during the optimum sowing period, using the recommended spacing and seed-rate.
5. Apply the recommended doses of fertilisers.
6. Protect the crop against insect pests, particularly aphids, by frequent spraying, using proper insecticides.

7. Harvest the crop in time to avoid loss owing to shattering, especially in the case of Raya.

Sesamum

Sesamum (*Sesamum indicicum* L.), also known as sesame, til and gingerly, is an important and ancient oil-yielding crop. Africa is considered to be the primary centre of origin of this crop because of the presence of its diverse wild species in that continent. India is the secondary centre of origin and another secondary centre is Japan.

Economic Importance

The sesamum seed is a rich source of edible oil. Its oil content generally varies from 46 to 52 per cent. Its grains may be eaten fried, mixed with sugar or in the form of sweetmeats. Sesamum oil is used as a cooking-oil in southern India. It is also used for anointing the body, for manufacturing perfumed oils and for medicinal purposes. Sesamum-cake is a rich source of protein, carbohydrates and mineral nutrients, such as calcium and phosphorus. The cake is edible and is eaten avidly by working classes. It is also a valuable and nutritious feed for milch cattle.

Botanical Description

The genus Sesamum belongs to the family Pedaliaceae which has a wide distribution, covering tropical Africa, Madagascar, Arabia, India, Sri Lanka, tropical Australia and a few of the eastern islands of the Malayan Archipelago.

Sesamum is an erect and annual herb. The plant is simple or branched. The leaves are opposite below and alternate above. The flowers are zygomorphic, solitary, occasionally two or three together, axillary, shortpedicelled, borne on the upper part of the stem or branches geniculate. The tube, bent downwords and is dialated above the oblique base. The calyx is small, and five sectioned, 0.5-0.6 cm long. The corolla is tubular, campanulate, approximately 3 cm long, with an applicate margin and slightly bilabiate, the upper lip is entire and the lower one is tripartite. There are four functional stamens and often one is sterile, didynamous. The antliers are sagittate. The ovary is superior, bilocular, but may be subdivided by false internal walls. The fruit is a capsule, erect and oblong. The capsule contains numerous small ovate seeds. The testa may be smooth or reticulate and may be white, yellow, reddish-brown or black. The seed is dicotyledonous, albuminous and oleaginous.

Distribution and Production

India, China Sudan, Mexico, Turkey, Burma and Pakistan are the important sesamum producing countries. India ranks first, both in the area and production of sesamum in the world. The annual area put under it in India is about 2-5 million hectares and the total production is nearly 52 thousand tonnes. Sesamum is grown on 21 lakh hectares in only eight states, viz. Uttar Pradesh (673,000), Rajasthan (562,000), Madhya Pradesh (345,000), Andhra Pradesh (237,000) Maharashtra (139,000), Gujarat (118,000), Tamil Nadu (117,000) and Orissa (103,000. Among other states only Karnataka has a sizable area (68,000) under sesamum. In the remaining states it is grown only on a small area and hence is a very minor crop there.

Climate and Soil

Sesamum grows in the plains and at elevations up to 1,200 m. It cannot stand frost, continuous heavy rain or prolonged drought. It is grown in sandy-loam to heavy black soils. Most of the crop is confined to lighter soils and its cultivation in heavy soils is limited to certain regions. The pH of sesamum growing soils ranges from 5.5 to 8.2. The sesamum growing soils are shallow to medium in Andhra Pradesh, Orissa, Andhra Pradesh and Tamil Nadu and are deep in Rajasthan, Gujarat and Maharashtra.

Mixed Cropping

sesamum is grown both as pure and mixed with other crops, whereas the semi-rabi and summer crops are taken as pure. The common component crops are red-gram (arhar) in Andhra Pradesh, bajra and red-gram in Gujarat, red-gram and jowar in Madhya Pradesh. cotton, groundnut and red-gram in Maharashtra, and black-gram, green-gram and horse-gram in Orissa.

Cultivation

In India, seasmum is grown in three seasons, viz. kharif, semi-rabi and summer. The kharif crop occupies over 70 per cent of the area, whereas the semi-rabi and summer crops occupy 20 per cent and 10 per cent area respectively. The kharif sesamum is sown in June-July with the onset of the monsoon and is harvested in December-January. The kharif and semi-rabi crops are entirely rainfed, whereas summer crop is grown under irrigation. The yield of the kharif crop is poor, whereas those of the semi-rabi and summer crops are high, as they are grown in rich soils and under better management. The preparatory cultivation for the kharif crop is usually not thorough.

For the semi-rabi and summer crops, the land is ploughed and harrowed repeatedly to secure a clean and fine seedbed. Except in Maharashtra and Gujarat, where line-sowing is practised, in all other states it is sown broadcast. The recommended spacings for the line-sown crop in different regions, however, range from 25 to 35 cm between the rows and 10 to 20 cm between the plants in the row. The seed being small, it is often mixed with sand to ensure its even distribution and the drill is operated rather shallow to avoid deep sowing. After sowing, the seed is covered lightly with a brush harrow. The seed-rate varies from 3 to 5 kg per hectare. The seed should be treated with Captan or Thiram at the rate of 3 g per kg of seed to control seed-borne diseases.

The rainfed crop is weeded and hoed once or twice and the irrigated crop is weeded often. Wherever the weed control with mechanical means is a problem, chemical weed control with pre-emergence application of Lasso at the rate of 3 litres/ha is recommended. The crop is generally recommended for different regions as 25-30 kg of W, 20-30 kg of P_2O_5 and 0-20 kg of K_2O per hectare. The crop is harvested when the leaves, stems and capsules begin to turn yellow and the lower leaves start shedding. To prevent the shedding of grains, the crop should not be allowed to become a drip in the field. The ripe plants are cut, carried to the threshing-yard, stacked for a week in the sun with the cut-ends downwards and well shaken or beaten to take out the grains from the dry capsules. Winnowing and cleaning completes the process.

Pests and Diseases

The leaf-eating caterpillar (*Antigastra catalaulmlis* Dup.) and the gall-fly (*Asphondylia sesalili* Felt) are the serious pests of sesamum. Two disulphur (0.5 kg/ha) sprays of Carbaryl 0.2%, or N on the 40th day and the 60th day, however give an elective control of both the pests. Stem and root-rot (*Macrophominia phaseoli* Maubl.) *Ashby phyllody* (virus, mycoplasina), bacterial leafspot (*Pseudomonas sesami* Matkoff) and leaf-curl are the important diseases of sesamum. Seed treatment with Ciptan at the rate of 3-9 Per kg of seed for controlling Macfphomila rot, 3 sprays of Agrimyein at the rate 01-100 ppm at 15-day intervals for controlling bacterial leaf spot are recommended. Both phyllody and leaf-curl can be controlled by spraying, the insecticides, such as Phosphomidon 0.05% or Dimethoate 0.05% or Dichlorovas 0.05%, which reduce the population of the insect vectors. The average grain yield of the kharif crop grown unmixed is 200 to 500 kg per hectare and the average yields of the semi-rabi crop and of the irrigated summer crop are 300 to 600 kg per hectare.

Varieties

The varieties differ not only in respect of the season of growth and the time of maturity but also in the number of capsules formed in the leaf axils, the habit of growth and the size, colour and oil content of the seeds. The seeds may be white, dull white, light brown dark brown, red or black and their oil content varies from 48-52%.

Hints for obtaining High Yields

1. A suitable variety must he chosen for the tract and the season.
2. The crop should be sown in lines.
3. The application of fertilisers, inter-cultivation and plant-protection measures must be taken up at the appropriate time.
4. Harvesting should be done after the crop matures but must not be delayed up to the stage when the capsules shed seeds.

Linseed

Linseed (*Linum uaitdissimum* L.) belongs to the family

Linaceae and the genus Linum which has 100 species. It has been grown from ancient times for fibre (flax) and for its seed which is rich in oil The oldest regions of cultivated flax are reported to be in Asia and on the Mediterranean coast.

Economic Importance

Linseed occupies a greater importance among oilseeds; owing to its various uses and special qualities. In India, it is grown mainly for seeds, used for extracting oil. The oil content of the seed varies from 33 to 47 per cent. Linseed oil is an excellent drying oil used in manufacturing paint and varnishes, oilcloth, waterproof fabrics and linoleum and as an edible oil in some areas. Linseed-cake is a very good manure and animal feed. Linseed straw produces fibre of good quality. Linseed is used in making paper and plastics.

Area and Distribution

Linseed is extensively grown in the countries of the temperate zone as well as in those of the tropical zone. The major linseed-growing countries are Argentina, the USSR, India, the USA, Canada, Pakistan and Australia. India accounts for about 1.9 million hectares, with a seed production of 4.98 lakhs of tonnes and occupies the third rank among the linseed-producing countries. Australia and Canada have the highest productivity of about 7 quintals per hectare, whereas India averages 255 kg per ha. This yield is the lowest in the world.

In India, Madhya Pradesh leads in yield and acreage, followed by Uttar Pradesh and Maharashtra, Bihar, Rajasthan, Karnataka and West Bengal also grow linseed in large areas. Madhya Pradesh and Uttar Pradesh together contribute to the national linseed production to the extent of about 70 per cent.

Climate and Soil

Linseed is grown in the range of latitudes between the 10th and 65th parallels, both north and south. Its cultivation is confined to low elevations, but it can he successfully grown up to 770 metres. Areas with the annual rainfall ranging from 45-75 cm are best suited for its cultivation. The seed crop does well under moderate cold, but the fibre crop grows best in cool moist climates. In India, the crop is grown in the rabi season from September-October to February-March.

Linseed can be grown on different kinds of soils, except the sandy and badly drained heavy clays or clay loams. It does well on the deep clayey black soils of central and peninsular India and on the alluvium loams of the Indo-Gangetic plains.

Cultivation

Fields of linseed are prepared as for other crops. Frequent ploughings are given to retain moisture and to check weeds in the field during the rainy season and a fine weed-free seedbed is prepared for sowing. BHC 10% or Aldrin or Chlordane 5% dust at the rate of 26-30 kg per ha is mixed within soil during the last ploughing before sowing to protect the seedlings against white-ants and the early damage from cutworms.

The main season for sowing linseed is October to November, depending upon the cessation of the rains. The crop is grown either broadcast or is drilled in lines 20 to 30 cm apart: In the case of broadcasting, which includes uterasowing also, the seed-rate is kept at 40 kg per ha. For line-sowing, the seed-rate varies from 20 to 510 kg per ha, depending upon the seed size. Thin sowing gives a good number of tillers, but for fibre and weed, thick sowing is desirable.

The application of fertilisers and irrigation increases the yield of the crop. Under rainfed conditions, the fertiliser dose including 30 kg of N and 15 kg of P per ha is given. If irrigation is available, two to three irrigations are sufficient. Under irrigated conditions, the linseed crop requires 60 kg of K and 30 kg of P per ha. The deep placement of fertiliser at sowing in the case of the rainfed crop gives better results. Under irrigated conditions, nitrogen is applied in two splits, half the dose as basal and the other half at the first irrigation, 40 to 50 days after sowing. The utera crop is fertilised at the rate of 10 kg of N per ha applied at the time of sowing. The rainfed crop generally requires no interculture. When the crop is irrigated, one weeding becomes essential.

Pests and Diseases

Linseed gall midge (*Dasyneura lini Barnes*) is the most wide spread and serious pest of linseed. The growing of early varieties and mixed cropping are reported to reduce the incidence of the pest. Linseed rust and wilt are the two most serious diseases. The growing of resistant varieties is recommended to control these diseases. The linseed crop starts maturing by the mid of February, depending on winter spread and sowing time. Plants turn golden yellow when the crop is mature and ready for harvesting. Harvesting is done when the crop is dead ripe with a sickle or by uprooting the plants. When the fibre is also desired along with seed, the harvesting of the crop is done at the stage of capsule maturity even when the crop is slight green. Linseed is threshed after it is completely dry. The crop is threshed either by beating it with a stick or by trampling it under the feet of the bullocks. Hand-beating is better to ge lien fibre, along with seed. Seed is separated from the chaff by winnowing and is stored in a dry place.

Yield

The average yield of a pure crop varies from 210 to 450 kg per ha of seed under rainfed cultivation. The crop in northern India generally gives higher yield than in central and peninsular India. The irrigated crop may yield 1,200 to 1,500 kg per ha. The percentage of oil in the seeds varies from 37 to 43, but the commercial extraction in expellers of rotaries gives about 33 per cent from small seeds and 34 to 36 per cent from hold seeds. Crushed in village ghanis, the seeds yield only 25 to 30 per cent of oil.

Varieties

In India, the linseed varieties are of two types:

(i) Peninsular types which have deep root-system.

(ii) Alluvial types which are shallow-rooted and are characterised by profuse tillering. Varieties differ in maturity, growth habit, and seed size and colour.

Hints for Getting Maximum Yields

(a) Clean, fine, weed-free seed-bed should be prepared by ploughing during the rainy season.

(b) 10% B.E.C. or 5% Aldrin or Chlordane at the rate of 25 to 30 kg/ha be mixed with the soil at the last ploughing before sowing.

(c) Early sowing escapes the attack of many diseases and pests.

(d) The application of 30 N+15 1 for the rainfed and 60 N+30 P for the irrigated crop and 10 N kg per ha for the utera sowing is recommended.

(e) Use the optimum seed rate of 30 to 40 kg per ha and a closer row spacing between 20 and 30 cm and 40 kg seed per ha for utera sowing.

(f) Cultivate the recommended varieties.

(g) Carry out plant-protection measures, if needed.

Castor

According to the available literature, castor (*Ricinus comrnunis* L.) is indigenous to Eastern Africa and most probably originated in Ethopia.

Economic Importance

It is generally grown for its, oil-yielding seeds. The oil content of the seeds varies from 35-58 per cent in different varieties, the average being about 47 per cent. Castor-oil is being used widely for various purposes. It is used as a lubricant in high-speed engines and aeroplanes, in the manufacture of soaps, transparent paper, printing-inks, varnishes, linoleum and plasticisers. It is also used for medicinal and lighting purposes. The cake is used as a manure and plant stalks as fuel or as thatching material or for preparing paper-pulp. In the silk-producing areas, leaves are fed to the silkworms.

Botanical Description

Castor belongs to the gellar Ricinus, its a member of the Euphorbiaccae which consists a vast number of plants mostly native to the tropics. The gellar Ricinus is considered to be momotypic and *R. comrnunis* as the only species which includes many polymorphic types. The cultivated types are dwarf annuals

The stem is erect, circular in section, partially, hollow, smooth, glabrous, with good branching. The stem is marked by a number of well-defined nodes, from each of which a leaf arises. Leaves are alternate, large, palniate with 5-11 lobes, acuminate, margins notchedy serrate or indented. They are carried on long stout petioles. The inflorescences are borne terminally on the main and lateral branches. Flowers are large, in terminal sub-panelled racemes, monoccious, apetalous, the upper portion of the raceme being occupied by the female flowers and the lower by the male flowers. The fruit is a roundish glaucous capsule, with three projecting sides covered with tough spines or smooth, three-loculed and three-seeded. Seeds are albuminous, anatropous, broad, oval, compressed with a marked carnnele and longitudinal raphe. The testa is thin, brittle, varying in colour and mottling. Below the testa is the thin tegmen, covering the whitish oily endospeim containing the embryo.

Distribution Area and Production

The main castor-growing countries are Brazil, India, USSR and Argentina. India ranks first in area (44,8000 hectares) but second in production (140,000 tonnes), Brazil being the, first. Andhra Pradesh (67.2%), Gujarat (12.7%) Karnataka (7.1%) and Orissa (5.8%) account for over 90% of the area and also production. In lighter soils, the land is repeatedly ploughed in summer, as and when the rains are received, and with the onset of monsoon rains the clods are crushed by working a country plough or harrow. In clay loams, only harrowing is done with a blade-harrow two to three times to bring the soil into proper tilth.

The seed is sown either in the plough furrow or with a seed-drill or by hand-dibbling. The seed-rate used for sowing behind the plough is 12 kg per hectare. For hand-dibbling, 8 to 10 kg of seed will be sufficient per hectare. Before sowing, castor seed should be treated with Thiram 3 g/kg of seed to prevent the attack of root-rot and Alternaria blight. The spacing adopted generally ranges from 90-120 cm between rows and 45-60 cm between plants in the row. One or two weedings and hoeings with bullock-drawn implements are given and sometimes, the plants are earthed up. The fertiliser dose recommended E for castor is 40 N-40 P-20 K kg/ha. The entire dose of P and K and half of W is applied basally and the remaining half of W about 30 days after sowing, as and when the soil moisture is adequate.

The improved varieties cultivated at present are early maturing and take about 150 to 180 days. When one or two capsules in a bunch show signs of drying, the whole cluster is generally removed and stacked, covered and weighed in a corner of the field or in a pit. The harvesting of unripe capsules in this manner has an adverse effect on the oil content of the seed. It is preferable to collect the fruits, as and when they ripen. When the whole crop is gathered, it is dried in the sun for a few days and the threshiug is done either by beating the dried capsules with a stick or by treading them under the feet of bullocks. Winnowing is done in the usual manner.

Pests and Diseases

The castor semi-looper (*Achow janaga* L.) and the capsule-borer (*Dicitocrocis punctiferalis*) are the two most serious pests of castor. Dusting BHC 10% in early stages or spraying 0.44% Ekalax or 0.1% Carbaryl or 0.05% 1 arathion on the crop will give an effective control on these pests.

The seedling blight and Alternaria blight Alternaria ricini (Yoshii) Hansford cause serious losses to this crop. The Sowing of the crop in low-lying and water-logged areas should be avoided to prevent the seedling blight from appearing. For alternaria blight, seed treatment with Thiram at the rate of 3 g/kg of seed is recommended.

Yield

The average yield of rainfed castor varies from 200 to 500 kg per hectare, that of the mixed crop from 100 to 200 kg and of the irrigated crop from 500 to 800 kg. In eleven years, the average total yield of castor in India has risen to 350 kg per ha.

Varieties

The varieties differ in the branching habit of the plants, colour of the stem and branches (red and green), the nature of capsules (smooth or shiny), duration (early or late) and the size of seed. The following are the principal improved varieties of castor recommended for cultivation in different states.

Hints for Getting Maximum Yields

1. Recommended variety for a particular area should he chosen.
2. Seed treatment with Thiram (3g/kg of seed) should he done before sowing to prevent the attack of root rot and alternaria blight.
3. Sowing should be done at the right time as recommended for different agro-climatic conditions. Sowing should he done with proper spacing as recommended for different regions.
4. The recommended fertiliser schedule for the region should he adopted.
5. Proper, timely and adequate plant-protection manure should be adopted to save the crop from insect pests and diseases.
6. Two to three intercultures should be done to provide aeration for the root-system and to keep the crop free from weeds.

7. The nipping of branches increases the size of the main spike and induces early maturity.
8. Harvesting must be done at the right time, as too early or too late harvesting results in reduced yield, oil and shattering of seed respectively.

Safflower

The three centres of origin, as suggested by Vavilov for safflower (*Carthamus tinctorius* L.), are India, Afghanistan and Ethopia. Decandole was of the opinion that Arabia was the most probable centre of origin.

The crop is now cultivated, primarily for its seeds which yield oil, though at one time it used to be grown for the extraction of a dye also. The seeds are edible and are eaten after roasting. Their oil content varies from 24 to 36 per cent, depending on the variety, soil, climate and other conditions. The cold-pressed oil is golden yellow and is used for culinary purposes, or for making soap. The oil obtained by dry hot distillation is black and sticky and is used only for greasing well ropes and leather goods exposed to water. Safflower oil has also good drying properties and is, therefore, used in the manufacture of paints, varnishes and linoleum. It can be mixed with white paint without any after-yellowing effects. The cake, particularly from decorticated seed, is used as a concentrated cattle feed, and that from undecorticated seed is sometimes used as a manure.

Botanical Description

The safflower plant is highly branched, herbaceous, thistle-like annual, varying in height from 30 to 160 cm. It has many branches each terminating in a flower. The inflorescence is broad, flat or slightly curved and densely bristled owing to the presence of numerous floral tracts.

The flowers are yellow to red, containing 20 individual ficrets, each of which produces a seed.

Distribution Area and Production

The important safflower growing countries, besides India, are the USA, Mexico, Ethopia, Spain, the USSR. and Australia. In India, it occupies 590,000 hectares with a production of nearly 130,000 tonnes. Over 98% of the area is concentrated in the states of Maharashtra (04.4%), Karnataka (26.0%) and Andhra Pradesh (8.0%).

Climate and Soil

The crop is grown during rabi primarily as a rainfed crop, but in some areas it is raised under irrigation. At all stages of growth of the crop, excessive rainfall or humidity increases damage from fungal diseases. Water-logging due to poor drainage or prolonged rains even for relatively short periods can cause substantial fall in seed-yield. Being drought-resistant, it is cultivated on all types of soil, including sandy soil, but it thrives best on water-retentive black soils and alluvial loams. The crop is fairly resistant to saline condition.

Cultivation

Safflower is grown mostly as a mixture with wheat, barley, gram and rabi jowar, three rows of safflower being planted after every nine, 12 or more rows of the main crop. Sometimes, it is sown as border rows surrounding the crops of wheat, rabi, jowar, etc., because being spiny, it protects the main crop against cattle trespass. As a pure crop, it is rotated with wheat, gram and rabi jowar. Occasionally, it is grown as a second crop after any quick-maturing kharif crop, e.g. green-gram (moong), black-gram (mash), groundnut, coriander, etc.

The number of ploughings should be as few as possible and be aimed at preserving the maximum amount of soil moisture. One or two ploughings with a country plough will be sufficient to remove the weeds and break the clods. Sowing is generally done in September and October. The seed-rate adopted varies from 5 to 12 kg per hectare, depending upon the soil fertility and the nature of the crop (pure or mixed). When grown in strips, a row spacing of 45 cm is adopted. Plants too close tend to have thinner stems or wear a superficial root-system, with a reduced number of flowers. In general, the pure crop is given one or two weedings, combined with hoeing on the 20th and the 45th day after sowing. The application of 20-40 kg of N per hectare results in a substantial increase in yield. The mixed crop shares the preparatory tillage, manuring and cultivation given to the main crop. When the plants have developed the central flowering head, they are topped to promote branching, flowering and seed production.

Harvesting and Yield

The crop matures in 4 or 5 months after sowing. The ripe plants are either cut or pulled out, stacked for a few days to dry, threshed by cutting with sticks, and the resulting material is winnowed to obtain clean seeds. The average yield of the pure crop is 400-500 kg of seed per hectare, whereas that of a mixed crop is about 100 kg.

Pests and Diseases

The safflower aphid (*Dactynotus carthami*), thrips (*FranklinieUa sulphurea*) and the flower bud fly (*Acanihiphilus helianthi* Rossi) are the serious pests of the crop. Spraying the crop with 0.1% Fenithion, 0.03% Dimethoate and 0.07% Endosulfan is recommended for controlling aphids, thrips and budfly, respectively.

Rust (*Puceinia carthami*), alternaria leaf-spot (*Alternai carthami*) and root-rot (*Phytophthora drechskri*) are the important diseases of safflower. The growing of resistant varieties is recommended to avoid damage from these

Hints for Obtaining the Maximum Yields

1. The sowing time should be adjusted in such a way that there will be no rainfall after flowering.
2. Nitrogen amounting to 20-45 kg per hectare has to be applied, preferably in two split doses, half at the time of sowing and the remaining half at flowering.
3. Timely inter-cultivation for the eradication of weeds and control measures against pests and diseases should be adopted.

Sunflower

The cultivated sunflower (*Helianthus annus* L.) is a native of southern United States and Mexico, from where it was taken to Spain before the, middle of the sixteenth century. In the nineteenth century, the cultivation of sunflower as an oilaced crop began in the Soviet Union and the majority of the present day varieties grown all-over the world trace back their origin to the USSR. The major sunflower-producing countries of the world are the Soviet Union, Argentina, Bulgaria, Pumania, Turkey and South America.

Economic Importance

Sunflower is mainly grown for its oil. The oil is used for culinary purposes, in preparation of vanaspati and in the manufacture of soaps and cosmetics. It is especially recommended for heart patients. Its cake is rich in protein and is used as a cattle and poultry feed.

Botanical Description

Sunflower belongs to the family Compositac and the genus Helianthus. The cultivated genotypes are characterised by a single stem terminating in a capitulum. Sunflower is protandrous, in which the male and female elements mature at different times. There appears to be a time-lag of 18-24 hours in the maturity of the male and female elements. So it is essentially a cross-pollinated plant, besides showing varying degrees of self-incompatibility.

Climatic and Soil Requirements

Sunflower is a day nuteral plant can be grown successgully in different seasons under varying conditions of day length, provided the temperature is favourable. Sunflower, basically a temperature plant, can also be grown commercially in tropical or sub-tropical conditions. It grows best with clearsky and occasionally rain shower during early stages. This corps requires a cool climate during germination and seedling growth and warm non-cloudy weather and high temperature during flowering to maturity.

Sunflower can tolerate temperature range of 8 to 30°C but the temperature below 16°C and above 40°C reduces yield and oil content. High temperature affects pollination and fertilisation, thus care should be taken to adjust the sowing date in such a way that flowering does not coincide with high temperature. The optimum temperature for best performance under controlled conditions is 27-28°C.

'Sunflower can be planted successfully when soil temperature exceeds 10°C, though germination begins at 3-5°C. Young plants (1 or 2 pairs of leaves) can withstand frost to -60C. However, plants are frost sensitive from 6-

8 leaf stage until flowering finishes. Late frost does not destroy the whole plant but destroys the fruiting head, leading to increased branching, resulting in small heads with many empty fruits. During flowering temperature should be between 18-20°C. Goods yields are obtained where the ruling temperature is 18-22°C showing that generally speaking the crop needs a warm climate.

Average minimum temperature below 17°C between flowering and maturity is necessary to ensure best oil quality. The crop is grown from 45°S to 55°N but greater production is between 20-50°S. The crop can be cultivated upto an altitude of 2500m but highest yield of oil is obtained below 1500m. Sunflower cannot tolerate frost and requires a frost free period of 120 days. It is resistant to drought but requires continuous availability of soil moisture for the best performance.

Sunflower can thrive on a wide range of soils such as sandy loam, black soil and alluvial soil, but it performs better on sandy loam soil than on clay soil under similar management. However, good drainage is more important than soil type. Thus, fertile, well drained, medium textured soils having heavy water holding capacity are best suited for the cultivation of sunflower. Sunflower is highly susceptible to water logging. Avoid cultivation of sunflower on soils affected by salts and with poor drainage, the optimum pH requirements 6.0 to 8.5.

Preparatory Tillage

Sunflower requires a fine seedbed, free of weeds and clods, for proper germination and good crop stand to bury the residues or previous crop, ploughing with disc harrow followed by 2-3 cultivations and plantings may be required fields vacated by potato may not require much seedbed preparation. However, adequate soil moisture is essential forgetting good germination.

Variety

MSFH8, PSFH-67,GKSFH-2002, Jwalamukhi, Mega-363 and NSFH-592 have been recommended. These hybrids are medium tall with dark green leaves their seeds are black in colour and mature in about 110-125 days mega-363 is a long duration hybrid and hence should not be selected for cotton areas. These hybrids have an average yield about 20 quintals per hectare with an oil content of 40.5 to 43.0 percent. In hybrids fresh seed is to be sown every time which adds to its cost of cultivation.

Sowing Time

Spring season is the best for getting higher and stable seed yield sunflower. Sunflower being a cross pollinated crop depends on honey bees for pollination and seed setting. Availability of honey bees is abundant during spring season due to favourable. Planting during spring should be made after the soil temperature exceeds 10°C at 10cm depth at 8 a.m. Earlier planting at lower temperature increases time of flowering without much earliness. Best results are obtained when crop is sown in the first fortnight of January which produces higher seed yield than later planting.

Sowing of short duration hybrids like GKSFH-2002,PSFH-67 and nsfh-592, however, can be extended upto second fortnight of January. Besides low yields, the late sowing may also face early mansoon rain at harvesting and threshing causing difficulty in its management.

Seed Rate and Seed Treatment

Use 5kg seed per hectare for hybrids to achieve the required plant stand. Low seed rate recommendation for hybrids is mainly because of smaller seed size and wider spacing for hybrids. Seed should be treated with Brassicol or Thiram @ 2 g per kg of seed to control seed and soil borne fungal disease.

Method of Sowing

A spacing of 60-30cm for hybrids has been found to be an optimum. These spacings result in better development of head, more seed filling and less stem breakage.

Crust Breaking

Rain after sowing, results in crust formation, which hinders the proper emergence of seedlings. Thus it may be necessary to break the crust, which can be done with harrowing. Harrow as soon as possible after the crust has performed and if possible, before seedlings have emerged.

Fertiliser Management

Fields studies have shown that hybrid sunflower responds well to nitrogen application upto 60 kg/ha. further increase in nitrogen results in lower seed and oil yield. Application phosphatic fertilisers @ 30kg P_2O_5/ha increase sunflower yield. Sunflower responds to K application only on soils deficient in available potassium and it is recommended to be applied @ 30 kg K_2O/ha on such oils.

Regarding application method, drilling the fertilisers besides the seed row has been found to be more beneficial for getting higher seed yield. Application of fertilisers below the seed was found to be injurious to germinating seeds and reduced seed emergence count and seed yield.

Entire dose of N, P and K should be applied at the same time of sowing. Prefer single superphosphate for P application as it also contains sulphur to which sunflower responds favourably. If single superphosphate is not available apply gypsum @ 25-30 kg/ha at sowing time. A supplementary dose of nitrogen @ 25-30 kg/ha may be top dressed a month after sowing sunflower if the crop exhibits symptoms of nitrogen deficiency.

Nitrogen deficiency appears as yellowing of the older leaves, which die prematurely. Leaves are also fewer and smaller. Plants with a severe deficiency are yellow and stunned, while young plants are of purplish colour. Plants are slender with longer internodes. The flower head is small with many unfilled seeds and the oil content is reduced.

Phosphorous deficiency results is delayed flowering, stunned growth and small leaves which prematurely die back from the leaf tip. Phosphorous deficiency is shown in small dark green leaves later turning pale with necrosis. Growth is restricted with short stiff stems, flowers are small, stunned malformed. Ripening is delayed, oil content and yield are reduced.

Potassium deficient plants are squat with a proportionately greater reduction in stem length then in leaf size. Initially leaves are dark green but turn yellow at the edges and die except the lower leaves. Potassium deficient plants have restricted growth with a rosette arrangement of the younger leaves. Later the leaves develop dark brown necrotic patches.

Water Management

Sunflower can withstand moderate drought because of its deep root system. However, it responds favourably to irrigation and its seed yield can be increased substantially by sustaining adequate moisture in soil by timely irrigations besides improvements in oil content.

Sunflower is highly moisture sensitive plant and its broad leaves exhibit wilting with slight moisture deficiency.

First irrigation to sunflower should be applied a month after sowing and interval during the subsequent months be reduced to 2-3 weeks during March and 8-10 days during the hots months of April and May. Irrigation should be stopped about 12-14 days before harvest. Sunflower is a tall growing plant with heavy head, therefore, irrigation at flowering and thereafter should be given on clam days to avoid lodging. The incidence of Sclerotinia stem-rot is more under irrigated conditions especially where the soil drainage is poor. Excessive and frequent irrigations on fine textured soils should be avoided.

Weed Management

Mechanical methods such as hoeing at 3-4 weeks after sowing followed by earthing up before flowering has been found to be effective in checking weeds. Use of tractor drawn implements before the crop attains 60-70 cm height has also been found effective. For better efficiency of manual weeding an improved 3 tine wheel hand-hoe could also be used.

Removal of weeds through mechanical/manual means quite effective but expensive. Moreover, often timely labour is not available. The studies initiated to find effective herbicides for weed control in sunflower, revealed that pendimethalin as pre-emergence application gave effective weed kill.

Harvesting and Threshing

The crop is ready for harvesting when the back of flower heads turn lemon yellow or yellowish brown near the stalk and the discs start drying. At this stage the seeds give a blakish look and are fooly ripe. The shadding of disc-flower residues from undamaged heads also indicate physiological maturity. Promptly harvesting seed after physiological maturity has many advantages; seed is least exposed to bird and rodent attack; the crop cycle is shorter resulting in more time to prepare for the next crop; reduces harvest seed losses; provides high quality seed and reduces exposures to bad weather.

The harvested heads are mildly dried and threshed by beating or trampling with tractors or sunflower threshers. The sunflower thershers are more efficient and economical. After threshing the produce should be thoroughly dried before storing otherwise in storage it gets affected by fungus and decomposes.

Allelopathic Effects of Sunflower

Researches show that sunflower depresses the growth and productivity of succeeding crops. The reason postulated are the allelochemicals being secreted by sunflower crop. Decomposing residues of sunflower could also bring about allelopathic effects. Phytotoxic compounds released into the soil rhizosphere have been observed to cause varying degree of allelopathy/ autotoxicity in crop.

Field studies conducted in the department of Agronomy, Punjab Agricultural University, Ludhiana on the allelopathic effect of spring sunflower on succeeding Kharif crops revealed that maize, sorghum, cowpea, pearlmillet, pigeonpea, mungbean and sesamum suffered 41.7, 36.7, 27.6, 28.8, 23.0, 10.8 and 9 percent reduction in yield, respectively, as compared to these when grown after wheat. Studies in allelopathic effects of different plant parts of sunflower indicated that sunflower roots reduced pigeonpea, cotton and mungbean yields the most followed by roots plus shoots treatment. Sunflower shoots did not have any harmful effect on growth of these crops.

Bioassay studies involving use of aqueous sunflower extract, decomposing plant residue and soil from area in proximity of sunflower crop revealed inhibition of gemination and growth of cotton, pigeonpea, sunflower and mungbean. These studies further showed that effect of leaf extract/residue was more than roots. The allelopathic effect of sunflower plant residue persisted upto 24 days after incorporation and beyond this period the effect was rather meagre.

Chromatograhic and physico-chemical analysis of allelochemicals in sunflower parts showed the presence of chlorogenic and isochlorogenic acids in both stems and leaves besides these some unidentified phytotoxins were also observed. The concentration of these was more in leaves than in stems. Further studies revealed that total phenolics was more in leaves followed by in roots and shoots. Sunflower raised under moisture stress conditions had higher phenolics as compared to the raised without any moisture stress.

Pests

Cut worm: It is polyphagous pest and cuts off the tender stalks, near ground, of young crop plants during night. During day time the caterpillars remain hidden in the soil. It is more serious during March-April when sunflower follows potato. When flat sowing is practised, apply 5 liters of Dursban/Radar 20 EC (Chlorpyriphos) per hectare before sowing. These insecticides should be mixed in 25 kg fine soil and broadcast uniformly after last ploughing but before planking. Sow crop on ridges to avoid cut worm damage in the early stages.

Tobacco caterpillar (Spodoptera litura): The polyphagus pest remain active throughout the year feeding on one or the other crop. The newly hatched caterpillar is pale-greenish with a black head and feeds on the outer epidermis of leaves. It poses a serous threat during April-May and defoliates the plants.

Cabbage Semi-looper (Plusia orichalcea): This insects feeds on many crops including the sunflower. The newly hatched caterpillars defoliate the crop and cause severe damage to it during April-May.

Bihar hairy caterpillar (Spilosoma obliqua): It is also called Bhabbu Kuta. This is highly polyphagous and voracious feeder. The nefarious pest causes severe damage and plants looke denuded. Newly hatched larva is bright brown and its body is covered with hair and legs are brown. At the older stage it is much darker and with dark brown hair of almost uniform length. These insects feed gregariously and should be collected along with the plant and buried deep in soil.

Head borer (Helicoverpa armigera): Head borer or American boll worm has become the noxious polyphagous pest in India and also causes heavy losses in sunflower. Newly hatched larva is green yellow which changes its colour with every instant.

The full grown caterpillar is brownish green having grey longitudinal streaks on it. The adults lay eggs on the flower buds. After hatching the larvae feed on developing seeds and make tunnels through them.

Head borer can be controlled by spraying 2.5 Kg of Sevin 50 WP (Carbaryl) or 2.0 litres of Asataf 75 SP (Acephate) or 2.125 litre Folothion 50 EC (Fenitrothion) or 2.5 litres of Dursban 20 EC (Chlorpyriphos) or Thiodan 35 EC (Endosulfan) or 1.25 litre of Nuvacron 30 SI (Monocrotophous) in 375 litres of water per hectare.

Repeat after two weeks if necessary. Endosulfan is comparatively safe for honey bees. Spray when honey bee activity is minimum i.e. early morning or late evening.

Diseases

Steam rot / cottony rot: It is soil and borne disease. Infestation of basal portion of stem takes place via mycelium that grows with the roots and the stem becomes soft and cotyledon leaves flatten on the ground. At bud formation soft decaying area appears on its stem and base and white mycelia appear later on. The stem becomes weak, leaves dry up and drop off with strong wind.

Infestation takes place on heads when fully developed and seed formation has taken place. The back of head becomes soft and turns brown. Fungus like cotton appears and seeds are also infested. The sclerotia of this pathogen, almost of the size of sunflower seed are produced in abundance which are visible in pith of rotton stem and also in the infested head. The affected tissue turns black and shredded. This disease can be effectively controlled by treating seed with thiram @ 2 g/kg seed before sowing.

Root rot or charcoal rot: It is soil borne disease that infects the roots. Roots turn brown and rot. Infection at early stages kills the seedlings rapidly. This also causes neck rot i.e. stem rotting immediately behind the head where infected stem turns brownish and darker towards the base. Roots turn grey and stunted. At later stages, it causes premature ripening, smaller and poor filled heads and a lower yield. For its control, treat the seed with Thiram @ 2 g / Kg seed and avoid moisture stress.

Head rot: High humidity and injury to plants favour infection at lower surface of head where it is attached to the stem. Irregular water soaked spots appear, which gradually enlarge and become soft. Shredding of back of heads takes place. Cracking of front side is also common. Seed filling is affected severely and seeds remain small and shrivelled. The yield losses range from 20 to 60 percent and oil reduction to the tune of 25-53 percent and increase in free fatty acids, which is undesirable. To avoid this disease plough the soil with furrow turning plough to bury the debris of previous crop and residues of sunflower; treat the seed with Thiram @ 2g/kg seed; avoid excessive stress of irrigation; grow recommended hybrid and don't grow sunflower repeatedly in the same field; Spray with Indofil M-45 @ 625 g per hectare in 250 litre of water at 15 days interval starting at about 60 days stage.

Utilisation

Sunflower crop has many uses:

- Edible oil of sunflower is of superior quality because of anticholestrol and non-cholestrol contents and is thus preferred over other oils for cooking food.
- Its oil is used for making vanaspati, soap and other allied products.
- Its oil is used in the manufacturing margarine.
- Sunflower meal in the form of cake is used for animal feed or manuring field.
- Hulls and stalks are used for the preparation of furfural.
- Green plants are used as forage or silage.
- Dry stalks are used as fuel.
- Pulp is also extract from plants.

— Sunflower husk finds use in the preparation of ethyl alcohol and yeast.

— Salted and fried kernels of sunflower are eaten as such and its meal is used for confectionery.

— The concentrates and isolates of defatted meals are used for protein additives, meat analogues, meat extenders, emulsifying agents, bakery products, beverages, snak products, etc.

— Seed is included in many commercial bird feeds.

Exploiting Male Sterility

Sunflower being a Cross-pollinated crop can be exploited for increasing the production of edible oil by developing hybrid sunflowers. The development of commercial hybrids is only possible by using male sterility, and fortunately such genetic systems are available in sunflower. Two types of male sterility occur in sunflower: (1) genetic male sterility linked with certain marker genes. This sterility is due to a recessive nuclear gene and linked to the green colouring matter in the seedling; (2) the cytoplasmic male sterility was first obtained from the interspecific cross H. petiolaria by Leetereq in France in 1965. This type is being widely used in developing hybrid sunflowers all over the world.

Niger

The Niger (*Guizotia abyssinica L.F. Cass.*) crop is said to be indigenous to tropical Africa and more specifically to Ethiopia.

Economic Importance

The crop is grown for its seed, used for extracting oil which is about 37 to 43 per cent of the seed weight. The oil is used for culinary purposes, for anointing the body, for manufacturing paints and soft soaps, for lighting and lubrication and for manufacturing cosmetics. The oilcake is a well-known cattle feed. The crop sheds a large quantity of dry leaves in the field and thereby adds organic matter to the soil.

Botanical Description

Niger belongs to the family Compositae. The plant is a stout, erect, smooth orscabrous leafy annual herb, about one metre tall. The leaves are opposite or the upper ones are alternate. The capitulum is peduncled, axillary and terminal heterogamous. Bay-florets are pistillate, scriate, fertile; ligule 2-3 toothed and yellow. Disc florets are hermaphrodite, fertile; limb campanulate and 5-fid. Achenes are glabrous, dorsally compressed, tip rounded, pappus absent.

Distribution, Area and Production

India is considered to be the chief Niger-producing country in the world, with an area of 4.8 lakh hectares, and with an annual estimated production of 1.2 lakh tonnes of seed. It is mainly grown in the states of Madhya Pradesh, Bihar, Maharashtra, Orissa and Tamil Nadu, of which the largest area of 2.38 lakh hectares is in Madhya Pradesh.

Climate and Soil

Niger is grown as a kharif crop under rainfed conditions. A moderate rainfall of 1,000-1,250 mm suits this crop and, as such, it is not cultivated in regions of heavy rainfall. Light red soils and brownish loams with sufficient depth and good texture are considered suitable for this crop. It is often grown on poor soils of coarse texture. It can be grown on well-drained heavy soils.

Cultivation

Niger is grown either as a pure crop or is grown mixed with minor millets. Two to three ploughings are enough to prepare the land for sowing. The field is made weed-free and levelled by planking before sowing. The crop is sown from June to August, either behind a country-plough or with a four-coultered drill. A spacing of 30 cm between lines and that of 10-15 cm between plants in the line are generally adopted. A seed-rate of 7-8 kg per hectare is adequate to maintain a good crop-stand and a pure crop.

About a fortnight after sowing, thinning is done to regulate spacing between plants. One or two hoeings are required for controlling weeds. In general, the Niger crop is not manured. But the application of 20 kg of N and 20 kg of P_2O_5 per hectare is recommended for obtaining higher yields of the crop. When sown as a mixed crop, Niger shares the manuring given to the main crop. The crop matures in November and December. The mature plants are cut with sickles at the base and heaped in the threshing-yard for a week. They are then spread in the sun to dry for two to three days, threshed by beating with sticks, and winnowed.

Yield

The average yield of Niger is about 200 kg/ ha, but under favourable conditions and with best management practices it yields up to 600 kg.

7

Soil Fertility Management

Fertilisers are compounds given to plants with the intention of promoting growth; they are usually applied either via the soil, for uptake by plant roots, or by foliar feeding, for uptake through leaves. Fertilisers can be organic (composed of organic matter, i. e. carbon based), or inorganic (containing simple, inorganic chemicals). They can be naturally-occurring compounds such as peat or mineral deposits, or manufactured through natural processes (such as composting) or chemical processes (such as the Haber process).

Plant Nutrition

The key-role of fertilisers and their judicious use in crop husbandry is well understood, when one is familiar with the general facts about plant nutrition. It is now known that at least 16 plant-food elements are necessary for the growth of green plants. These plant-nutrients are called essential elements. In the absence of any one of these essential elements, a plant fails to complete its life cycle, though the disorder caused can, however, be corrected by the addition of that element. These 16 elements are: Carbon (C), hydrogen (H), oxygen (O), nitrogen (N), phosphorous (P), sulphur (S), potassium (K), calsium

(Ca), magnesium (Mg), iron (Fe), manganese (Mn), zinc (Zn), copper (Cu), molybdenum (Mb), boron (B) and chlorine (Cl).

Green plants obtain carbon from carbon dioxide from the air; oxygen and hydrogen from water, whereas the remaining elements are taken from the soil. Based on their relative amounts, normally found in plants, the plant nutrients are termed as macronutrients, if large amounts are involved, and micronutrients, if only traces are involved. The micronutrients essential for plant growth are iron, manganese, copper, zinc, boron, molybdenum, and chlorine. All other essential elements listed above are macronutrients.

Most of the plant nutrients, besides carbon, hydrogen and oxygen, originate from the soil. The soil system is viewed by the soil scientists as a triple-phased system of solid, liquid and a gaseous phases. These phases are physically separable. The plant nutrients are based in the solid phase and their usual pathway to the plant system is through the surrounding liquid phase, the soil solution and then to the plant root and plant cells. This pathway may be written in the form of an equation as:

$$\text{M (Solid)} \rightarrow \text{M (Solution)} \rightarrow \text{N (Plant root)} \rightarrow \text{(Plant top)}$$

where 'M' is the plant nutrient element in continual movement through the soil-plant system.

The operation of the above system is dependent on the solar energy through photosynthesis and metabolic activities. This is however, an oversimplified statement for gaining a physical concept of the natural phenomenon, but one should bear in mind that there are many physico and physico-chemical processes influencing the reactions in the pathway. The actual transfer in nature takes place through the charged ions, the usual form in which plant-food elements occur in solutions (liquid phase of the system). Plant roots take up plant-food elements from the soil in these ionic forms. The positively charged ions are called 'cations' which include potassium (K^+), Calcium (Ca^{++}), magnesium (Mg^{++}), iron (Fe^{+++}), zinc (Zn^{++}), and so on. The negatively charged ions are called anions and the important plant nutrients taken in this form include nitrogen (NO^{-3}), phosphorous (H_2PO^{-4}), sulphur (SO^{-4}), Chlorine (Cl), etc.

The process of nutrient uptake by plants refers to the transfer of the nutrient ions across the soil root interfaces into the plant cell. The energy for the process is provided by the metabolic activity of the plant and in its absence no absorption of nutrients take place. Nutrient absorption involves the phenomenon of ion exchange. The root surface, like soil, carries a negative charge and exhibits cation-exchange property. The most efficient absorption of the plant nutrients takes place on the younger tissues of the roots, capable of growth and elongation.

In this respect, root-systems are known to vary from crop to crop. Hence their feeding power differs. The extent and the spread of the effective root-system determines the soil volume trapped in the feeding-zone of the crop plant. This is indeed an important information in a given soil-plant system which helps us to choose fertilisers and fertiliser-use practices. The absorption mechanisms of the crop plants are fairly known now. There are three mechanisms in operation in the soil-water-plant systems. They are:

(i) the contact exchange and root interception,

(ii) the mass flow or convection, and

(iii) diffusion.

In the case of contact exchange and root interception, the exchangeable nutrients ions from the clay-humus colloids migrates directly to the root surface through contact exchange when plant roots come into contact with the soil solids. Nutrient absorption through this mechanism is, however, insignificant as most of the plant nutrients occur in the soil solutions. Scientists have found that plant roots actually grow to come into contact with only 3 percent of the soil volume exploited by the root mass, and the nutrient uptake through root interception is even still less.

The second mechanism is mass flow or convection, which is considered to be the important mode of nutrient uptake. This mechanism relates to nutrient mobility with the movement of soil water towards the root surface where absorption through the roots takes place along with water. Some are called mobile nutrients. Others which move only a few millimetres are called immobile nutrients. Nutrient ions such as nitrate, chloride and sulphate, are not absorbed by the soil colloids and are mainly in solution. Such nutrient ions are absorbed by the roots along with soil water.

The nutrient uptake through this mechanism is directly related to the amount of water used by the plants (transpiration). It may, however, be mentioned that the exchangeable nutrient cations and anions other than nitrate, chloride and sulphate, which are absorbed on soil colloids are in equilibrium with the soil solution do not move freely with water when it is absorbed by the plant roots.

These considerations, therefore, bring out that there are large differences in the transport and root absorption of various ion through the mechanism of mass flow. Mass flow is, however, responsible for supplying the root with much of the plant needs for nitrogen, calcium and magnesium, when present in high concentrations in the soil solution, but does not do so in the case of phosphorous or potassium. The nutrient uptake through mass flow is largely dependent on the moisture status of the soil and is highly influenced by the soil physical properties controlling the movement of soil water.

The third mechanism is diffusion. It is an important phenomenon by which ions in the soil medium move from a point of higher concentration to a point of lower concentration. In other words, the mechanism enables the movement of the nutrients ion without the movement of water. The amount of nutrient-ion movement in this case is dependent on the ion-concentration gradient and transport pathways which, in turn, are highly influenced by the content of soil water. This mechanism is predomionant in supplying most of the phosphorous and potassium to plant roots.

It is important to note that the rhizophere volume of soil in the immediate neighbourhood of the effective plant root receives plant nutrients continuously to be delivered to the roots by diffusion. However, when the nutrient concentration builds up far excess of the plant in the reverse direction. These are some of the choice of fertilisers and fertiliser practices for practising scientific agriculture.

Functions of Plant nutrients

The plants require, the following essential nutrien for their normal development:

Carbon	Nitrogen	Calci
Hydrogen	Phosphorous	M esium
Oxygen	Potassium	phur
Iron	Zinc	Chlorine
Manganese	Boron	Copper
Molybdenum		

Carbon is obtained from carbon dioxide of the air; Oxygen from air and water; Hydrogen from water; Nitrogen from air and soil or both, and all other nutrients from the soil. Soil is a the most important source of plant food.

Nitrogen, Phosphorous and Potassium are known as primary plant nutrients; Calcium, Magnesium and Sulphur are secondary nutrients; Iron, Manganes, Copper, Zinc, Boron, Molybdenum and Chlorine as trace elements or micronutrients. The primary nutrients and secondary nutrients elements are known as major elements. This classification is based on their relative abundance, and not their relative importance. The micronutrients are required in small quantities, but they are as important as the major elements in plant nutrients.

Air is the primary source of Nitrogen for plant nutrient. Only leguminous crops can directly use this free Nitrogen with the help of symbiotic bacteria of the genus Rhizobium. Other plant derive from soil their Nitrogen in the form of Nitrogen and Ammonium. Nitrogen and Ammonium are produced in the soil by action of microorganisms on the soil organic matter. Non symbiotic microorganisms can fix free Nitrogen of the air and make it available to plant in Ammonium and Nitrates forms.

Nitrogen encourages the vegetative development of plan[illegible] by importing a healthy green colour to the leaves. It also [illegible]ntrols, to some extent the efficient utilisation of phospho[illegible]us and Potassium. Its dependency retards growth and [illegible]oot development, turns the foilage yellowish or pale green, [illegible]stens maturity, causes the shrivelling of grains and low[illegible]s crop yield. The older leaves are affected first. An [illegible]ess of Nitrogen produces leathery (sometimes crinkled, [illegible]ark-green leaves and succulent growth. It also delays [illegible]maturation of plants, impairs the quality of crops [illegible] barley, potato, tobacco, sugarcane, and fruits; increases susceptibility to diseases and causes 'lodging' of cereal crops by inducing an undue lengthening of the stem internodes.

Phosphorous influences the vigour of plants and improves the quality of crops. It encourages the formation of new cells, promotes root growth (particularly the development of fibrous roots), and hastens leaf development through emergence of ears, the formation of grains, and the maturation of crops. It also increases resistance to diseases and strengthens the stems of cereal plants, thus reducing their tendency to lodge. It offsets the harmful effects of excess nitrogen in the plant. When applied to leguminous crops, it hastens and encourages the development of nitrogen-fixing nodule bacteria.

If phosphorous is deficient in the soil, plants fail to make a quick start, do not develop a satisfactory root-system, remain stunted and sometimes develop a tendency to show a reddish or purplish discolouration of the stem and foilage owing to an abnormal increase in the sugar content and the formation of anthoscyanin. However, the deficiency of this element is not easily recognised as that of nitrogen. It has also been observed that cattle feeding on the produce of deficient soils become dwarfed, develop stiff joints and lose the velvetty feel of the skin. Such animals show an abnormal craving for eating bones and even soil itself.

Potassium enhances the ability of plants to resist diseases, insect attacks, and cold and other adverse conditions. It plays an essential part in the formation of starch and in the production and translocation of sugars, and is thus of special value to carbohydrate-rich crops, e. g. sugarcane, potato and sugar-beet. The increased production of starch and sugar in legumes fertilised with potash benefits the symbiotic bacteria and thus enhances the fixation of nitrogen. It also improves the quality of tobacco, citrus etc. With an adequate supply of potash, cereals produce plump grains and strong straws. But an excess of element tends to delay maturity, though, not to be the same extent as nitrogen.

Plants can make up and store potassium in much larger for correcting zinc deficiency. The symptoms of zinc deficiency appear generally in younger leaves, starting with intervienal chlorosis leading to a reduction in shoot growth and the shortening of internodes. Mottle leaf, little leaf, etc. in the case of trees are symptoms of zinc deficiency. The buds of several deficient maize plants become white; in citrous interveinal chlorosis and mottled leaf occur. In calcareous soils and in soils with very high phosphorus content, zinc deficiency is commonly expected to occur.

The principal function of zinc in plants is as a metal activator of enzymes. In highly weathered coarse textured soils zinc deficiency appears under an intensive cropping programme. The availability of zinc is least between pH 5. 5 and 7, but its availability increases at a lower pH. At a higher pH above 7, zinc availability becomes a complex problem, as the positively charged zinc ion gets converted into a negatively charged zincate complex whose availability tends to be reduced in alkaline soils. When the calcium ion is predominent, the highly insoluble calcium zincate is formed and zinc availability gets seriously limited. The application of soluble zinc salts or zinc chelates to the soil is generally recommended to correct its deficiency. Foliar sprays are advocated, especially for orchard trees for amending zinc deficiency. About 5 to 50 kg of zinc sulphate per hectare is used for such purposes.

The symptoms of *boron* deficiency vary with the kind and age of the plant, the conditions of growth and the severity of the deffiency. Each crop produces its characteristic growth abnormalities associated with boron deficiency, such as yellows and rosetting in lucerne, snakehead in wallnuts, die-back and corking of fruits in apple, corking and pitting of fruits in tomatoes, hollow stem and the bronzing of curd in cauliflower, the brown-heart diseases in table-beets, turnips, etc.

Molybdenum deficiency produces whip-tail in cauliflower, broccoli and other *Brassica* crops. The deficiency of this element reduces the activity of the symbiotic and non-symbiotic nitrogen-fixing microorganisms.

It was in 1954 that *chlorine* was proved to be an essential micronutrient. Its deficiency under field conditions has not been reported so far. In water-culture solutions, the leaves of chlorosis, necrosis and an unusual bronze discolouration on tomatoes.

Sodium is not an essential element for plant growth. But some crops, such as beet, celery, cabbage, kale, knol-khol, radish, rape and turnip, benefit greatly by application of soluble sodium salts, specially if the soil is deficient in potassium. Sodium is also of direct benefit to plants indigenous to the sea-shore or to irrigated arid regions. Salts of this element are said to release more of potassium from the exchange complex and to help to maintain phosphorus in a more available form. They also serve as a partial substitute for potassium in the case of potatoes and cotton.

Maintenance of Soil Fertility

No two soils are alike either in respect of their nature or in respect of quantities of plant nutrients they contain. Under a given situation, the system of farming, soil management and manuring practices, etc. , influence the productivity of soils and crop yields obtained from them. The quantities of the three primary nutrients- N, P_2O_5 and K_2O removed from a hectare of land by some of the important crop are shown in table 1.

Table 1. The quantities of plant nutrients removed from soil by different crops (Kg/ha)

Crop	*Yield (grain)kg/ha*	N	P_2O_5	K_2O
Rice	2, 240	34	22	67
Wheat	1, 568	56	24	67
Jawar	1, 792	56	15	146
Bajra	1, 120	36	22	66
Maize	2, 016	36	20	39
Barley	1, 120	41	20	35
Sugarcane	67, 200	90	17	202
Groundnut	1, 904	78	22	45
Mustard	672	22	11	28
Linseed	1, 008	19	12	33
Cotton	448	30	17	45
Jute	1, 568	67	34	67
Tea	896	45	13	28
Coffee	896	34	11	34
Tobacco	1, 456	94	57	91

It is estimated that the different agricultural crops in India remove about 4. 27 million tonnes of nitrogen, 2. 13 million tonnes of phosphoric acid, 7. 42 million tonnes of potash and 4. 88 million tonnes of lime per year. The production of larger yields through improved varieties of crops and intensive cultivation will increase the depletion of nutrients still further. But erosion and leaching cause additional losses. The present production of synthetic nitrogenous fertilisers in the country reached 1. 5 million tonnes of nitrogen in 1975-76, and the bulky organic manures might supply another 1. 5 million tonnes of total nitrogen.

The amounts of phosphorous, potash, etc. , added to the soil are very small. It is thus obvious that the current huge drain on nutrient supplies will continue to impoverish the soils unless these supplies are replenished by natural or by artificial means, the principal methods of supplementing natural recuperation and for improving the productive capacity of the soils are:

(i) to add organic matter to the soil, so that through decay, it may furnish a more or less continuous supply of nutrients for crops, and

(ii) to restore or increase the amount of defficient nutrients by the application of fertilisers.

The urgent need of the constantly expanding agriculture production to meet the requirements of the continually increasing human and cattle populations in India makes the supply of additional plant nutrients through fertilisers and organic manures, a problem of supreme importance.

Types of Manures and Fertilisers

Indian soils are usually very poor in organic matter as well as in nitrogen. Pohsphate deficiency is less widespread and potash deficiency generally occurs in compact areas. In acid soils the addition of lime steps up production. Materials which are commonly used to maintain and improve soil fertility may be classified as follows:

(1) *Manures.* These are relatively bulky materials, such as animal or green manures, which are added mainly to improve the physical condition of the soil, to replenish and keep up its humus status, to

maintain the optimum conditions for the activities of soil microorganisms and make good a small part of the plant nutrients removed by crops or otherwise lost through leaching and soil erosion. They, thus, supply practically all the elements of fertility which crops require, though not in adequate proportions. The plant-food elements contained in a manure are released in an available form after it is applied to the soil and is decomposed by soil microorganisms. Similarly, the green manures add not only substantial amounts of organic matter but also nitrogen.

(2) *Fertilisers*. Fertilisers are inorganic materials of a concentrated nature; they are applied mainly to increase the supply of one or more of the essential nutrients, e. g. nitrogen, phosphorous and potash. Fertilisers contain these elements in the form of soluble of readily available chemical compounds. This distinction is, however, not very rigid. In common parlance the fertilisers are sometimes called 'chemical', 'artificial' or 'inorganic' manures.

(3) *Concentrated Organic Manures*. Some of the concentrated materials, such as oil-cakes, bone-meal, urine and blood are of organic origin. The use of manures and fertilisers is complementary and not as a substitute for each other.

(4) *Bulky Organic Manures*. The properties and role of organic matter and humus in the soil have been explained already.

Farmyard Manure

Good-quality farmyard manure is perhaps the most valuable organic matter applied to a soil. It is the most commonly used organic manure in India. It consists of a mixture of cattle dung, the bedding, used in the stable and of any ramnants of straw and plant stalks fed to cattle. Though its crop-increasing value has been recognised from time immemorial, more than 50 per cent of the cattle dung produced in the country today is burnt as fuel and is thus lost to agriculture.

Not only this tremendous waste, but also the tradition method of preparing and storing the farmyard manure is generally faulty. The cattle-dung, together with stable-waste and house sweeping, is first collected in the open backyard, and when a cartload has been collected, it is removed to another heap or to an uncovered pit in a common plot outside the village. The loose heaps lie exposed to the sun, with the result that the raw organic matter dries up quickly and does not rot properly. Very often, a part of the dry dung is blown off by wind or washed away by rain. Cattle urine is either not conserved or is stored in a defective manner.

American studies on the distribution of soil derived elements between urine and faeces of dry cows have shown that 95 per cent of Potassium, 63 per cent of nitrogen and 50 per cent of sulphur are contained in the urine. The wastage of nitrogen-rich urine, the loss of nitrogen (in the form of ammonia) due to the fermentation of exposed cattle dung, and the washing away of soluble mineral elements be leaching reduce its manurial value in India to a great extent. Its average content of plant nutrients under Indian and European conditions is shown below for comparison:

Percentage content:

	N	P_2O_5	K_2O
India	0. 3	0. 15	0. 3
European countries	1. 0	0. 30	1. 0

About half of this nitrogen, one-sixth of phosphorous and more than half of potash are readily soluble and subject to dissipation. However, the loss of nitrogen and minerals elements caused by careless handling can be reduced greatly by using absorbent bedding for cattle, storing dung in stone or brick-line pits, mixing large quantities of straw and other vegetable matter with cattle dung, and keeping the heap compact and moist. Thus, if urine is properly conserved, the loss of soluble mineral elements through seepage is prevented, bacterial decomposition of raw organic matter is encouraged, plant nutrients are made soluble, and nitrogen losses are minimised.

If urine is not conserved in the bedding used for cattle it must be collected in covered *pucca* cistern and then, added to the dung in the manure pit. Nitrogen in the urine is mainly in the form of urea, which readily changes into the highly volatile ammonium carbonate through bacterial action, and quickly loses ammonia thereafter by evaporation.

The pit may be 1 m in depth, 1. 3 to 1. 5 m in width and 4. 5 to 6 m in length, depending upon the number of cattle on a farm. The filling of the pit should be 'sectional' and when each section of three or 1. 3 m in length is filled to about 45 cm above the ground level, it should be clustered with 2. 5 cm layer of a mixture of mud and dung in equal proportions. Before plastering, 4 to 5 buckets of water should be added to the manure in the pit. Plastering conserves moisture and nitrogen and also prevents housefly nuisance. The manure becomes ready for use in about 4 to 5 months after plastering.

The quality of manure is also improved by the concentrated feeds given to cattle. Cotton-seed, cotton-seed cake, linseed-meal, wheat bran, grain husk, groundnut cake, gram, horse-gram, etc. are rich in nitrogen, phosphorous, potassium, magnesium and sulphur. It has been found that in the case of adult working-cattle about 80 per cent of nitrogen and the other mineral elements contained in the feed is recovered in urine, faeces and other animal by-products. Accordingly, manure from cattle fed on cereal straws and grass hay is much less valuable than that from animals fed on legume hays, grains and concentrates.

In foreign countries, considerable attention has been given to the use of preservatives on manure. Calcium sulphate or gypsum and superphosphate have proved most promising in preventing the escape of ammonia. Gypsum has been found specially effective as an ammonia-absorbing agent. Superphosphate, besides absorbing ammonia, supplies additional phosphorous and, thus, improves the crop producing capacity of the manure.

Partially rotten farmyard manure should generally be applied to the soil about three to four weeks before the sowing of a crop. In case there is sufficient moisture in the soil, there sill be enough time for its decomposition and for improving the soil structure. Its application too long before sowing a crop will either cause a drying up of the rotted manure or too quick decomposition, depending on the incidence of rains. But in each case, there will be a serious loss of ammonia and nitrogen. If the manure is already well rotted, it is advisable to apply it just before sowing to a crop.

This procedure is particularly essential in the case of light soils. In any case, after the manure is carted to the field, it should be evenly spread and worked into the soil soon to avoid the loss of nitrogen. The existing practice of leaving the manure in small heaps scattered in the field for several days, before spreading it in the field and incorporating it into the soil, results in the serious deterioration of its quality, particularly if strong winds blow. In vegetable and fruit cultivation, the application of well-rotten manure, in conjunction with fertilisers to young plants individually, has been found to give the best results. In Egypt, even the cotton crop, which is invariably grown on ridges, is given a top-dressing a handful of the rotten manure being applied to the soil at the base of each plant before an irrigation, and is worked into the soil with a hand hoe.

The practice of penning cattle, sheep and goats in the fields in summer is common in some parts of the country. Folding 7, 000 sheep for one night is said to add the equivalent of 149. 3 quintals (14. 93 tonnes) of cattle dung. The fresh dung left in the field in such cases rapidly dries up. This drying checks ammonification and loss of nitrogen. With the first fall of rain, the dung is worked into the soil. It, therefore, does not lose much of its fertilising value. Further more, its beneficia effects on the physical condition of the soil are undeniable. However, sheep-folding is said to make the land more weedy.

There must be adequate moisture in the soil for the proper decomposition of organic matter. Farmyard manure can, therefore, be applied to all crops grown in the rainy season or grown under irrigation. The quantity of manure to be applied to unirrigated crops varies from 1. 5 to 2 cartloads per hectare in areas of heavier rainfall.

If sufficient farmy manure is not available, it may be applied at the usual rate to a part of the land, say, to one-third or one-forth of the area, in rotation every year, so that all parts of the field receive the manure regularly once in three or four years. For irrigated field crops, the rate varies from 10 to 20 cartloads. Sugarcane, maize and garden crops, such as potatoes, turmeric, ginger, vegetables and fruits receive still higher doses, amounting sometimes to 15 to 25 cartloads. A cartload of manure, measuring 9 cubic metres, weighs about half a tonne.

It must be stressed that the value of farmyard manure in soil improvement is due to its content of principal nutritive elements and its ability to

(i) improve the soil tilth and aeration,

(ii) increase the water-holding capacity of the soil, and

(iii) stimulate the activity of microorganisms that make the plant-food elements in the soil readily available to crops.

The supply of organic matter, which is later converted into humus is a property of farmyard manure.

The use of farmyard manure alone causes an imbalance in nutrition owing to its relatively low content of phosphoric acid. Therefore, to keep the soils well supplied with all the essential elements of plant food in a readily available form, and also to keep them in good 'heart', it is advisable to use the bulky organic manures in conjuction with superphosphate and such other artificial fertilisers as contain the particular plant food or foods in which a soil may be deficient or which the crop to be grown may specially require.

Composted Manure

Another method of augmenting the supplies of organic is the preparation of compost from farmhouse, and cattle-shed wastes of all types. Composting has been advocated and adopted extensively during the past 25 years. Composting is the process of reducing vegetable and animal refuse (rural or urban) to a quickly utilisable condition for improving the maintaining soil fertility.

Research conducted in India and abroad has shown that good organic manure similar in appearance and fertilising value to cattle manure similar in appearance and fertilising value to cattle manure can be produced from waste materials of various kinds, such as cereal straws, crop stubble, cotton stalks, groundnut husk, farm weed and grasses, leaves, leaf-mould, house refuse, wood ashes, litter, urine-soaked earth from cattle-sheds and other similar substances. These raw vegetables materials are rich in cellulose and other readily decomposable carbohydrates and have a carbon-nitrogen ratio of 40 or more to 1.

The direct application of such undecomposed, low nitrogen organic matter as manure brings about a temporary deficiency of mineral nutrients (specially nitrates and ammonium compounds) in the soil by stimulating the growth of micro-organisms, which in turn, compete with crop plants for available nitrogen, phosphorous and other elements. Hence, before using them as manure, it is necessary to compost or partially decompose them. This process lowers the carbon-nitrogen ratio to about 10 or 12 to 1.

Two methods of composting waste organic materials are usually recommended. One depends on aerobic and other one unaerobic decomposition. In both cases, the farm wastes have to be used as a bedding for cattle in order to absorb a large of the animal urine. In the aerobic process, the used bedding, the sweepings from cattle sheds and some urine-soaked earth from the stable floar are removed everyday, mixed with a little cattle dung and 2 or 3 handfuls of wood ashes are deposited on a well drained site to gradually builded up a low pile about 30 or 45 cm in height, about 5 m in width and of any convenient length. The pile is built up before to start of rainy season. After the first heavy shower, the waited material in 1. 2 strip on each side of the long heap is turn with a rake on to 2. 4 m wide stripe in the middle, so raising the height of the heap to nearly 1 m. This process prevents loss of misture and ensures a quick start of decomposition.

When the heap sinks appreciably and such a sinking takes about 3 to 4 weeks, it is given a turning and made into fresh heap, thus mixing the outside material with that from inside. After about a month or more, depending on incidence of rains the heap is given a final turning on a cloudy or moderately rainy day and rebuilt the vacant part of the original position. The composed becomes ready for use in about 4 months. This method is eminently suited for composting in the rainy season. The following proposition of raw materials is considered suitable for compost making:

Parts by weight:

Mixed farm residues and cattle-shed	400
Urine-soaked earth	56
Fresh cattle dung	60
Wood ashes	6

In Tamil Nadu, the application of 90 kg of partially fermented dung in the form of a thin suspension, with 22. 50 kg of bone-meal per tonne of dry material to be composted is recommended.

If urine-soaked earth and cattle-dung are not available, the raw organic materials can still be composted, provided more than one-third of the residues is soft and finely broken, such as fallen leaves, leaf-mould, kitchen wastes, grass clippings, green and succulent weeds, plant trimming and clopped wheat, barley and soft cereal straw. The use of ordinary soil and wood ashes or lime is, however, essential.

In the anaerobic process, the mixed farm residues are collected in pits of a convenient size. Each day's collection is spread in a thin layer, sprinkled with a mixture of fresh cowdung (4. 5 kg), ashes (140 to 170 g) and water (18 to 22 litres) and compacted. The pit is filled till the raw material stands 38 to 46 cm above its edge, and then is then plastered with a 2-5 cm layer of a mixture of mud cowdung. Under such conditions, decompostion is anaerobic and high temperatures do not develop. Insoluble nitrogen compounds gradually become soluble and the carbonaceous matter is broken down into carbon dioxide and water.

The loss of ammonia is negligible, because in high concentrations of carbon dioxide, ammonium carbonate is stable. The plastered pit also prevents the fly nuisance. The well-made compost contains 0. 8 to 1 per cent of nitrogen and has all the good properties of farmyard manure. It can be used in the same way as the latter. The anaerobic process is particularly suited for use by gardeners in or near cities and towns.

Composting Methods

Raw Materials

The materials are needed are mixed plant residues, animal dung and urine, earth, wood ash and water. All vegetables wastes available on a farm like weeds, stalks, stems, fallen purnings, chaff, fodder remnants, green matter and so on, are collected and stacked in a pile.

Hard woody material like cotton or pigeon-pad stalks and stubble are first spread on the farm road and crushed under Vehicles viz. tractors or bullock carts before being piled. Such hard materials should in any case not exceed

10 percent of the total plants residues. Green material which are soft and sufficient and allowed to wilt for 2 or 3 days to remove access moistures before stacking; they tend to pack closely, if they are stacked in the fresh state. While stacking, each material is spread in layers of about 15cm thickness until the heap is about 1m and 50cm high. The mixture of different kinds of vegetable residues ensures a more efficient decomposition.

The heap is then cut into vertical slices and about 20-25 kg is put under the feet of cattle in the shed as bedding for the night. The next morning the bedding, along with the dung and the urine, and urine earth, is taken to the pits where the composting is to be done.

Pit Method

This method includes following steps :

— *Site and size of pit*: The site selected for the compost pit should be near to cattle shed and water source at high level so that no rain water gets in during the mansoon season. A temporary shed may be constructed over it to protect the compost from heavy rainfall. The pit should be about 1 m deep, 1. 5-2. 0 m wide and of any suitable length.

— *Filling the pit*: The material brought from the cattle shed is spread and on each layer is spread a slurry of dung made with 4. 5 kg urine earth and 4. 5 kg of inoculum taken from a 15-day-old composting pit. A sufficient quantity of water (nearly 90%) is sprinkled over the material in the pit to the wet it. The pit is filled in this way layer-by-layer and it should not take longer than 1 week to fill. Care should be taken to avoid compacting the material in any way.

— *Turning*: The material is turned 3 times during the whole period of composting (i) after 15 days from filling the pit, (ii) another 15 days, (iii) after another 30 days. At each turning the material is mixed thoroughly, moistened with water and replaced with the pit.

Heap Method

During rainy seasons or in regions with heavy rainfall the compost may be prepared in heaps above ground. When sufficient nitrogenous material is not available a green manure or leguminous crop like sunhemp is grown on the fermenting heap by sowing seeds after the first turning. The green matter is then turned in at the second mixing.

Forming the heap

The heap usually commenced with a 20 cm layer of carbonceous material such as leaves, hay-straw, sawdust, wood chips and chopped corn stalks. This is then covered with the 10 cm of nitrogenous material such as fresh grass, weeds or garden plant residues, garbage, fresh or dry manure or digested sewage sludge. The pattern of 20 cm carbonaceous material and 10 cm nitrogenous material is followed until the pile is 1. 5 m high and they are normally wetted so that they feel damp but not soggy. The pile is sometimes covered with soil or hay to retain heat and is turned at 6- and 12-weeks-interval. In the republic of korea, heaps are covered with thin plastic sheets to retain heat and it has also resulted in the death of insects.

If materials are limited, the alternate layers can be added as they become available. Also, all materials may be mixed together in the pile, if one is careful to maintain the proper proportions. Shredding the material speeds up decomposition considerably; most materials can be shredded by running over it several times with a rotary-mower.

Advantages and limitations

Preparation on a large scale can be done through community composting. There is lack of protection from rain and wind. A considerable amount of water is needed ans so the heap method is not suitable in near areas of scanty rainfall.

The intense aerobic decomposition to which the material is subjected no doubt shortens the period of composting but it leads to heavy losses of organic matter and nitrogen. Therefore the C:N ratio should be maintained between 30 and 40 to reduce such losses.

Bangalore Method

Preparation of the pit

Trenches or pits 1m deep are dug; the breadth and length of the trenches can be made depending on the availability of land and the type of material to be composted. The selection of site for 1 pit is made as mentioned in the Indore method. The trenches should preferably have sloping walls and a floor of 90 cm slope to prevent waterlogging.

Filling the pit

Organic residues and night soil are put in alternate layers and after filling, the pit is covered with a 15-20 cm thick layer of refuse. The materials are allowed to remain in the pit without turning and watering for 90 days. During this period the material settles down due to reduction in volume of the biomass and additional nightsoil and refuse in alternate layers are placed on top and plastered or covered with mud or earth to prevent loss of moisture and breeding of files. The material undergoes anaerobic decomposition at a very slow rate and it takes about 180-240 days to obtain the finished product.

Advantages and limitations

The recovery of the finished product is greater as compared to aerobic composting but loss of nitrogen is negligible. Labour requirements is less than for the Indore method as turning of material is not done; labour is needed only for digging and filling the pits.

The methods requires along time to produce a finished compost and so takes up more land use. A uniform high temperature is not assured in the biomass. Problems of odour and fly breeding need to be attended too.

Synthetic Compost

In the preparation of synthetic compost, the organic nitrogen as dung, required by microorganism, can be completely substituted by inorganic nitrogen compounds like ammonium sulphate and urea which are utilised equally effectively for decomposition of carbonaceous materials into compost. The process of preparations of synthetic compost developed by Hutuchinson and richards is based on this principle. This facilitates the utilisation of large quantities of various organic waste material where supplies of dung are either short of the requirement or not available at all, as on mechanised farms.

The basic principle of C:N ratio in manure preparation can be applied to add nitrogenous fertilisers in sufficient quantity to decompose. The material to be composted is moistened. This is sprinkled with the fertiliser solution and then with lime. Superphosphate may be added to fortify the phosphorous content of the manure. The treatment is continued layer-wise until the heap or pit is filled to the size and allowed to ferment. The manure becomes ready for application in about 120-180 days and resembles with farmyard manure in its action on soil and plant growth.

Leaf Compost

Leaf composting, can be achieved by heap or ditch composting or by windrow composting. Windrow are preferred as they allow efficient handling of materials Provide good aeration, allow sufficient of water and are easy to be performed.

It is suggested the formation of uniform-shaped windrow from 2. 40 m^{-3}. 60 m at the base and 2. 40m^{-3} high and of any convenient length. Windrow built too high will have excessive compaction at the base resulting in anaerobic conditions. Windrows built too low will not allow sufficient insulation to sustain thermophilic temperatures during cold weather. To ensure proper aeration, it is important to break apart tightly compacted leaves.

Though reduction in size may aid in rapid decomposition, it is not desirable in leaves because it increases the compaction making more frequent aeration necessary. When the incoming leaves are not adequately moist, it is desirable to add water to maintain a proper moisture regime of 40 to 60%.

The C:N ratio of leaves is relatively high. It can be as high as 80, and needs to be amended with nitrogen. Sewage sludge, urea and grass clippings are good sources of nitrogen. If a nitrogen source is to be added, caution should be exercised to distribute it uniformly throughout the windrow, lest it may result in undesirable anaerobic conditions and uneven decomposition. Proper aeration can be maintained by periodical mixing of the material.

Under optimum environmental composting conditions, leaf compost will be ready between 180 and 270 days. Leaf mould will have final PH range of 6-7. It is suggested the use of finished compost (leaf mould) as converting material (10-15cm) in the subsequent preparation of leaf compost to supply a heavy inoculum of microorganisms.

Enrichment With Phosphorous

Phosphorous enriched compost is prepared by adding 5% superphosphate at the filling of the compost pits. Other sources of phosphorous for this purpose are powered rock phosphate, preferably of low grade (less than 11% P), can be used with profit. Besides, phosphorous it is a source of calcium and micronutrients. Bonemeal will provide nitrogen as well as phosphorous; it contain 9-11% P and 2-4% N. Steamed bonemeal is more easily ground than the fresh material. It contains a little less nitrogen but more phosphorous than the raw material. Basic slag provides calcium, magnesium and trace nutrients and small amount of phosphorous. Banana residues contains about 1-1. 5% phosphorous on an ash basis.

Enrichment With Potassium

Granite dust of powdered potassium-containing minerals like feldspars can be added to enrich compost. Potassium and other deficient elements can be added to compost by including plant materials which contain appreciable amounts of those elements. Water hyacinth for example, is a rich source of potassium and of many other elements required by plants. Banana skin and stalks contain 34-42% potassium on an ash basis, seaweeds are rich in iodine, boron, copper, magnesium, calcium and phosphorous. Leaves are also a good source of trace elements and should form a part of every compost heap. Potato peel is rich in trace elements and dry potato vines contain 1% potassium, 4% calcium and 1% magnesium.

Vermi Composting

Vermi-Composting is the use of earthworms for composting of organic residues. Earthworms can consume practically all kinds of organic matter. One worm, weighs about 0. 5 to 0. 6 g, wastes as their own body weight per day and produces cast of the same weight per day. It is estimated that 1000 tonnes of moist organic matter can be converted by earthworms into 300 tonnes of compost. Organic materials undergo complex biochemical changes in the intestines and vermi-composting is an appropriate technique for disposal of non-toxic solid and liquid organic wastes. It helps in cost effective and efficient recyclicing of animal wastes (poultry, equine, piggery excreta and cattle dung) agriculture residue and industrial wastes using low energy, excreta together with their cocoons and undigested feed make up vermicasting.

The castings of earthworms are rich in nutrients (N, P, K, Ca and Mg), and also in bacterial and actinomycetes population. The actinomycetes population in worm cast is over 6 times more than in the original soil. The introduction of worms into compost heap has been found to mix the materials, aerate the heaps and hasten decomposition. Turning the heaps is not necessary, if earthworms are present to do the mixing and aeration. Besides rural and urban wastes, effluents from agro-industries viz dairies, tanneries, pulp and paper mills, distilleries etc. can be treated by using earthworms.

Benefits of Earthworms

Earthworms help in the preparation of compost maintaining soil health as follows :

1. Improvement in fertility of soil.
2. Arnelioration of physical condition of soil.
3. Mixing of sub-soil and top soil.
4. Correction of undetermined deficiences in plants.
5. Use of earthworms in recycling of city and rural wastes, sewage waste waters and sludge, and industrial wastes e.g.. paper, food and wood industries.
6. Supplementing traditional feeds

Earthworm can be divided as surface living (epigeic) and burrowing (epianecic) worms. Epigeic or compost worms are found on surface and are reddish brown e.g.. Lumbricus rubellus (red worms). Of many species of earthworms tested for mass culture all over the world, Eisenia fetida, Eudrilus eugeniae and Perionyx excavatus come in the above order of preference for their ability to compost organic wastes.

A bedding material is compounded from miscellaneous organic residues saw dust, cereal straw, rice husks, sugarcane trash, bagasse, paper, cardboard, coir waste, grasses etc. and is moistened well with water. The wet mixture is stored for 30 days covered with a damp sack and is thoroughly mixed several times. When fermentation is complete, chicken manure and green matter e.g.. Leucaena leaves or water hyacinth is added. The material is placed in the boxes in the ans sufficiently loose for the worms to burrow and should be able to retain moisture. The proportion of the different materials will vary according to the nature of the material but a final nitrogen content of about 2. 4% should be aimed at. A pH value as near neutral as possible is necessary and the boxes should be kept at temperatures between 20°C and 27°C.

At higher temperatures the worms will aestivate and at lower temperatures they hibernate. For each 0. 1 m^2 of surface area 100 g of breeder worms are added to the boxes. In spite of their being able to eat the bedding material, the worm at this stage are regularly fed @ 1 kg of feed a day for every kg of worms. The feed stuffs used are again various types of organic matter and include partially digested cowdung, chicken manure, Leucaena leaves, vegetable waste and water hyacinth. Some form of protection against predators like birds, vats, ants, frogs, leeches and centipeded is provided to the worms.

Vermi-composting in pits

Vermi-composting is done in pits and in vitro. Both of these are discussed here.

Bamboo poles are laid in paralled row on the pit's. Its floor is with a lattice of wood strips. Necessary drainage is provided because worms can not survive in a waterlogged condition. Alternatively to this and sand can be placed in the bottom of the pit to facilitate proper drainage. Above this a thick layer (15-20 cm) of good loamy soil should be spread. The pit can now be filled with available organic residues such as animal manure, leaves and green weeds, crop residues etc. Moisture levels of the contents of pit is maintained through addition of required amount of water. The worms from breeding boxes are introduced in the organic refuse, the worms immediately burrow down into the damp soil.

The compost pit is left for 60 days. It should be shaded from hot sunshine and it must be kept moist. Within 60 days about 10 kg of castinge would have been producer per kg of worms. The pit is then excavated to an extent of about two-thirds to three-quarters and the bulk of the worms removed by hand or by serving. This leaves sufficient worms in the pit for further composting and the pit can be refilled with fresh organic residues and continued.

The compost can be sun-dried and sieve to give a good quality compost. The average nutrient content of vermi-compost is N 0. 6-1. 20%, P_2O_5 1. 34-2. 20%, K_2O 0. 4-0. 67%, CaO 0. 44% and MgO 0. 15%.

The excess worms that have been harvested from the pit can be used in the other pits, sold to other farmers for compost inoculation, and may be used as animal and poultry feed or fish food.

Method of pit vermi-composting:

— *Selection of earthworm* : Earthworm which is native to the local soil and vermi-compost may be used.

— *Size of pit* : Any convenient dimensions may be prepared. This can hold 20, 000-40, 000 worms giving 1 tonne manure/month (30 days).

— *Preparation of vermibed* : a 15-20 cm thick layer of good loamy soil above a thin layer (5cm) of broken bricks and sand should be made. This layer is inhabited by earthworms.

— *Inocluation of earthworms* : About 100 earthworms are introduced as an optimum inoculating density into a compost pit provided with a vernibed.

— *Organic layering* : It is done on the vermibed with fresh cattle dung. The compost pit is then layered to about 5 cm with dry leaves or hay. Moisture content of the pit without flooding is maintained through the addition of water.

— *Wet organic layering*: It is done after 28 days with moist/green organic waste which can be spread over it to a thickness of 5 cm. This practice can be repeated every 3-4 days. Mixing of wastes periodically without disturbing the vermibed ensures proper vermi-composting. Wet layering with organic waste can be repeated till the compost pit is nearly full.

— *Harvesting of vermi-compost*: At maturation, the moisture content is brought down by stopping the addition of water for 3-4 days. This ensures drying of compost and migration of worms into the vermibed. The mature compost, a fine loose granular mass is removed out from the pit, dried and packed.

— *Rate of application*: Mature vermi-compost is recommended @ 0. 5 tonnes/ha.

To boost vermi-compost production following suggestions should be followed :

(i) A mixture of cattle, sheep, horse dung with gram and wheat bran and vegetable wastes forms the ideal feed for worms.

(ii) Mixing of gram bran with dung mixture in 3:10 ratio increase the biomass.

(iii) Mixing of wheat bran to dung mixture in 3:10 ratio hastens the growth of worms. Addition of kitchen waste in the same proportion increases the worm population.

(iv) The biogas sludge and poultry dropping in equal quantities enhance the worm population and the biomass.

In-vitro vermi-composting

This is also called as bioconversion in soil. This involves the application of the basal dose (5 tonnes/ha) of vermicastings and covering with 2. 5 cm. Layer of organic matter followed by 10 cm layer of sugarcane trash, crop residues or city swastes. The worms hatch out within 10 days.

Vermi-composting of agricultural wastes

With the aim of vermi-composting of agricultural wastes an experiment was conducted at the Indian Agricultural Research Institute, New Delhi. The usefulness and efficiency of earthworms in composting of agricultural residues was studied.

Two trials were conducted with mixed organic materials. Trial A, had more of green materials like grasses and Leucanena leaves mixed with soil and paper. Experiment B had 4 kg of composting materials consisting of 2 kg of paddy straw, 1 kg soil 500 g twigs, and 500 g shredded paper and was laid in layers per pit. The bottom most layer was laid with twigs to allow for percolation of excess moisture. The compost pit was kept moist and care was taken to avoid waterlogging. After 10 days of initial decomposition, worms (*Eisenia fetida*) obtain from M/s Biogenic Ltd. , Mumbai were introduced @ 100 worms/pit. After the worms burrowed into the deeper layers, the pit was covered with a thin layer of soil.

The results of both of the trials show that introduction of worms in the organic matter piles was found helpful in composting. Organic carbon was reduced at different intervals of composting. In the inoculated compost, nitrogen content was appreciably augmented and the C:N ratio was narrowed to a desirable level. The worms were also active in decomposition of high C:N ratio of compost was brought to 51 only whereas with worms it was narrowed down to 21. The use of worms also increased the available phosphorous content of compost. Conjoint use of cellulolytic fungi and earthworms showed better result then other of them alone.

Town compost

In recent years, large-scale composting of town refuse and night-soil in properly constructed trenches away from human habitations has been taken up successfully by the municipalities of many large and small towns. Trenches, 1 to 1. 2 m wide, 75 cm deep and of convenient length, are filled with successive layers of night-soil, town refuse and earth, in this order. The compost gets ready in about three months. The following figures of volume-weight conversion will be found useful in the preparation of town compost.

Volume	Weight in kg or q
1 cu. m of refuse	= 318 kg or 3. 18 q
1 litre of night-soil	= 0. 991kg
1 cu. m of compost	= 636 kg or 6. 36 q
1 cartload of refuse	= 95. 40 kg

Social prejudice against the use of this valuable compost has disappeared and town-composting is almost being rapidly adopted in other localities all over India. With a suitable modifications, such as the provision of trench latrines, it can be taken up in villages too. Particulars of the process may be obtained from the National Extension Service Officers.

Sewage and sludge

The liquid waste and sewage contain large quantities of

plant nutrients and are used for growing of sugarcane, vegetables and fodder crops near many large towns by operating sewage-farms. In many places, the undiluted sullage has been found to be too strong for healthy plant growth and if it contains readily oxidised organic matter, its use actually reduces nitrates present in the soil. The disadvantages are still greater if sewage is used on land without preliminary treatment.

The soil quickly becomes 'sewage sick' owing to the mechanical clogging by colloidal matter in the sewage and the development of anaerobic organisms which not only reduce the nitrates already present in the soil but also produce alkalinity. Bacterial contamination makes the eating of raw vegetables on untreated sewage a real danger to health.

For these reasons, it is now usual to construct a setting or a septic tank in which sewage is stand allowed to relieve it of the heavier portion of the solid matter in it, or to undergo a preliminary fermentation. The effluent from the settling-tank, however, still carries a large amount of objectionable colloidal matter, and the deposit of sludge that settles in the tank is of small manurial value and is often offensive. These defects can be removed by thoroughly aerating the sewage in the settling-tank by blowing air through it.

The sludge that settles at the bottom in this process is called 'activated sludge' It has the remarkable property of bringing about the rapid oxidation of the organic matter present in fresh sewage. It is also in offensive and on dry-weight basis, contains 3 to 6 per cent N, about 2 percent P_2O_5 and 1 per cent K_2O in forms that can become readily available when applied to the soil. Similarly, the effluent is a clear, odourless liquid containing nitrates in solution, from which most of the pathogenic bacteria originally present have been removed. Both the activated slidge and

teysmanni is grown to provide green leafy twigs for manuring paddy.

For the proper rotting of the green manure, it is necessary that the green material should be succulent and there should be adequate moisture in the soil. Plant at the flowering stage, contain the greatest bulk of succulent organic matter with a low carbon/nitrogen ratio. The incorporation of the green-manure crop into the soil at this stage allows a quick liberation of nitrogen in the available form. With advancing age, the percentage of carbonaceous matter in the plants increases and that of nitrogen decreases. If the material with a side carbon/nitrogen ratio is ploughed under, microorganisms bring about its decomposition, draw upon the released nitrogen and mineral nutrients and cause a temporary nutrient deficiency.

Sometimes *methi* or *dhaincha* is sown in-between the rows of the newly planted sugarcane crop, or cluster-bean is planted between the rows of irrigated American cotton. When the leguminous plants are five to six weeks old, they are incorporated into the soil. Sugarcane and cotton are claimed to benefit from the legumes. The decay of green matter in this case seems to occur when it best serves as a fertiliser for the beneficiary crop.

The increase of yield after green-manuring is usually of the order of 30 to 50 per cent. The fertilising value of the legume crop can be increased a great deal by manuring it with superphosphate. This practice not only increases the phosphorous content of the green-manure plants, but also encourages their plant growth on the whole, thus converting an inorganic fertiliser into an organic manure. Green manures have a marked residual effect also.

Fertilisers

Despite the special steps taken in recent years to increase the supply of farmyard and other bulky organic manures, the available quantities are insufficient to meet the existing and prospective needs. Fertilisers have also the advantage of smaller bulk, the resultant easy transport, relatively quick availability of their plant-food constituents and the possibility of their application in proportions suited to the actual requirements of different crops and soils. Fertilisers are usually classified according to the three principal elements and may, therefore, be included in more than one group.

Nitrogenous Fertilisers

According to the manner in which their nitrogen is combined with other elements, the nitrogenous fertilisers are divided into four groups; nitrate, ammonia and ammonium salts, chemical compounds containing nitrogen in the amide form, and plant and animal by-products.

Sodium Nitrates

It is also known as 'Chilean' nitrate. It owes its importance as a pioneer nitrogenous fertiliser. It occurs in natural deposits in northern Chile and is refined before shipping. The refined product contains about 16 per cent nitrogen in the nitrate form, which renders it directly available to plants. For this reason, it is highly valued as a source of nitrogen when applied as top-and side-dressings, especially to young plants and garden vegetables, which need readily available nitrogen for quick growth leached out from the soil. For wheat, maize, barley, cotton, sugarcane, etc. , it is as beneficial as ammonium sulphate. Sodium nitrate is particularly useful for acidic soils. Its continued and abundant use in soils is said to cause deflocculation and develop a bad physical condition in regions of low rainfall. It should be stored in a dry ware-house

Ammonium Sulphate

It is the most widely used fertiliser in the country. It is a white crystalline salt, containing 20 to 21 percent ammoniacal nitrogen. It is easy to handle and it stores well under dry conditions. During the rainy season, it sometimes, forms lumps. These lumps should be powedered before use. Being soluble in water, it acts quickly, but despite its high solubility, its nitrogen is not readily lost in drainage, because the ammonium ion is retained by the soil particles. It is, therefore, very suitable for wet-land crops, e. g. paddy and jute.

It has also been found useful on wheat, cotton, sugarcane, potatoes and many other crops grown on a wide variety of soils. It has, however, an acid effects on the soils. Its long-continued use increases soil acidity and lowers the yield. The application of this fertiliser to acid soils improved the yield of tea plants considerably. It is advisable to use this fertiliser in conjunction with bulky organic manures to safeguard against the ill effects of continued application of ammonium sulphate to field and horticultural crops. Ammonium sulphate can be applied before sowing, at sowing time, or as a top-dressing to the growing crop. It should not be applied along with, or too close to, the seed, because in concentrated form, it affects seed germination very adversely.

Ammonium Nitrate

It is a white crystalline salts, containing 33 to 35 per cent nitrogen, half as nitrate nitrogen and half in the ammonium form. In the ammonium form, it cannot be easily leached from the soil. This fertiliser is quick-acting, but highly hygroscopic and not fit for storage. Granulation of the material and a light coating of the granules with oil reduce hygroscopicity to some extent. It has an acidulating effect on the soil. Under certain conditions, it is explosive, and, therefore, it should be handled cautiously.

'Nitro Chalk' is the trade name of a product formed by mixing ammonium nitrate with about 40 per cent lime-stone or dolomite. It is granulated, non-hazardous and less hygroscopic. It contains 20. 5 per cent nitrogen, half in the form of ammonia and half as nitrate. The presence of lime in it makes it particularly useful for acid soils.

Ammonium Sulphate Nitrate

It is a mixture of ammonium nitrate and ammonium sulphate. It is available in a white crystalline form or as dirty-white granules. This fertiliser contains 26 per cent nitrogen, three-fourths of it in the ammoniacal form and the rest (6. 5 per cent) as nitrate nitrogen. It is non-explosive and not as deliquescent as ammonium nitrate. It is readily soluble in water and is very quick-acting. Its keeping quality is good and it is useful for all crops. Its keeping quality is good and it is useful for all crops. Its acid effect on the soils is only one-half of that of ammonium sulphate. It can be applied before sowing, at sowing time or as a top-dressing, but it should not be applied along the seed.

Ammonium Chloride

It is a white crystalline compound, possessing a good physical condition and containing 26 per cent

the effluent can be used with safety for manuring and irrigating crops. However, under no circumstances should any produce grown on a sewage-farm be eaten uncooked.

Night-soil or poudrette

Very few towns India are equipped with complete sewage. The sanitary disposal of night-soil with an effective control of foul smell and fly nuisance is, therefore, a serious problem all over the country. Since human excrements are a potential source of soil improvement, public health authorities in many towns make the necessary arrangements for its conservation and conversion into a form in which it can be safely used as a manure. The dehydration of night soil, as such, or after admixture with absorbing materials, e. g. soil, ash. charcoal and sawdust, produces a poudrette that can be easily used as a manure.

The mixing of night-soil with an equal volume of ash and 10 per cent powdered charcoal produces an odourless material, containing 1. 32 per cent potash and 24. 2 per cent time. The addition of 40 to 50 per cent of sawdust to the night-soil yields straightway a dry, acidic poudrette which may contain 2 or 3 per cent nitrogen. The annual potential quantity of manurial ingredients in the night-soil from India's present population of 600 millions is estimated at approximately 8. 1 million tonnes of dry matter containing almost 0. 4 million tonnes of nitrogen, 0. 25 million tonnes of phosphoric acid and 0. 17 million tonnes of potash.

Green manures

Despite special efforts at increasing the supplies of farmyard manure and compost, the supply of farm and other organic manures is scarce and ever becoming more costly. Green-manuring, wherever feasible, is the principal supplementary means of adding organic matter to the soil. It consists in the growing of a quick-growing crop and ploughing it under to incorporate it into the soil.

The green-manure crop supplies organic matter as well as additional nitrogen particularly if it is a legume crop, which has the ability to acquire nitrogen from the air with the help of its root-nodule bacteria. A leguminous crop producing 8 to 25 tonnes of green matter per hectare will add about 60 to 90 kg of nitrogen when ploughed under. This amount would equal an application of three to ten tonnes of farmyard manure on the basis of organic matter and its nitrogen contribution. The green manure crops also exercise a protective action against erosion and leaching.

The crops most commonly used for green-manuring in this country are the following:

Sunnhemp *(Orotalaria juncea)*, dhaincha *(Sesbania aculeata)*, cluster-bean *(Cyamopsis tetragonoloba)*, senji *(Melilotus parviflora)*, cowpea *(Vigna catjang, V. sinensis)*, horse-gram *(Dolichos biflorus)*, pillipesara *(Phaseolus trilobus)*, berseem or Egyptian clover *(Trifolium alexandrinum)*. Lentil *(Lens esculenta)* is recommended in Kashmir for green-manuring paddy.

Sown in late autumn, it is said to provide a winter cover and make new growth in early spring for ploughing under before the sowing of paddy. Sunnhemp is the most outstanding green-manure crop. It is well suited to almost all parts of the country and fits in well with sugarcane, potatoes, garden crops and the second-season paddy in southern India and with irrigated wheat in the north. *Dhaincha* is in wide use in Assam, Bengal and Tamil Nadu. It does well on alkaline and water-

ammoniacal nitrogen. It is extensively used on paddy in Japan, In India, it is used largely in industries. In general, it is similar to ammonium sulphate in action. It is usually not recommended for tomatoes, tobacco and such other crops as may be injured by chlorine.

Urea

It is a white, crystalline, organic chemical. It is a highly concentrated nitrogenous fertiliser, containing 45 to 46 per cent of non-proteined organic nitrogen. It is fairly hygroscopic and presents considerable difficulty, it is also produced in granular or pellet forms and is coated with a non-hygroscopic inert material. It is highly soluble in water and , therefore, subject to rapid leaching. It is, however, quick-acting. When applied to the soil, its nitrogen is rapidly changed into ammonia. Like ammonium nitrate, urea supplies nothing but nitrogen. It may be applied at sowing time or as a top-dressing, but should not be allowed to come into contact with the seed. It is suitable for most crops and can be applied to all soils.

Ammonia

It is a gas containing about 80 percent of nitrogen. Under suitable conditions of temperatures and pressure, it becomes liquid (anhydrous ammonia). Another form, 'aqueous ammonia', results from the absorption of ammonia gas into water, in which it is soluble. Ammonia is used as a fertiliser in both these forms. Anhydrous ammonia can be applied by introducing it into irrigation water, or directly into the soil from special containers, which makes its use rather expensive. Its possibilities as manure for paddy, sugarcane and cotton are being investigated at Bangalore in the Karnataka state. In manurial experiments on cotton in Maharashtra, aqueous ammonia has been found to be as efficient as ammonium sulphate.

Calcium Ammonium Nitrate

Calcium ammonium nitrate is a fine free-flowing, light brown or grey granular fertiliser. It is commercially prepared from ammonium nitrate and ground limestone. It is almost neutral and can be safely applied even to acid soils. Its total nitrogen content may vary from 25 to 28 per cent. Half of this total nitrogen is in the ammoniacal form and half is in nitrate form. According to the prescribed standards, its moisture content should not be more than 0. 5% by weight. As regards the particle size, 90 per cent of the material should pass through 4 mm IS sieve and be retained on a 1-mm IS sieve but not more than g per cent shall be below 1 mm.

Organic nitrogenous fertilisers

These fertilisers include plant and animal by-products, such as oil cakes, fish manure and dried blood from slaughter houses. Before their organic nitrogen can be used by the crops, it is converted through bacterial action into readily usable ammonia-nitrogen and nitrate nitrogen. These fertilisers are, therefore, relatively slow-acting, but they supply available nitrogen for a longer period. Furthermore, they may also small amounts of organic stimulants that they may contain, or of some of the minor elements needed by plant.

Oil-cakes of different kinds are produced in India to the tune of about two million tonnes annually. They contain not only nitrogen but also some phosphoric and potash, besides a large quantity of organic matter. In addition to the three fertilising constituents (N, P_2O_5 and K_2O), the oil-cakes invariably contain 2 to 15 per cent of oil, depending on whether the oil is extracted by using solvent process or with expellers, hydraulic presses or indigenous *ghanis*. This residual oil, however, in practice, affect their manurial value. Owing to the great importance of using edible oil-cakes as a cattle feed, their utilisation as fertilisers is undesirable. They should be fed to cattle and the excrements may be used as manure. Inedible cakes, like castor cake, *neem, mahua* cake and *karanj* cake can, however, be recommended for use in conjunction with quicker-acting chemical fertilisers.

Dried blood or blood-meal contains 10 to 12 per cent highly available nitrogen and 1 to 2 per cent phosphoric acid. It is a very quick-acting manure and effective on all crops and all types of soils. It should be used in the same way as oil cakes. Fish manure is available either as dried fish or as fish-meal or powder. In regions where fish oil is extracted, the residue can be used as a manure. Depending on the type of fish, its manurial constituents vary from 5 to 8 per cent of organic nitrogen and from 4 to 6 per cent of phosphoric acid. It is quick-acting and suitable for all crops and soils. It should preferably be powdered before use.

Phosphate fertilisers

These are classified as natural phosphates, treated or processed phosphates, and by-product phosphates and chemical phosphates.

Rock Phosphate

It occurs as natural deposits of rock in Morocco, the United States of America, Poland, Russia, Tunisia, Algeria, Algeria, Brazil, Egypt, Nehru and some islands in the Pacific Ocean and the Indian Ocean. It contains 25 to 35 per cent phosphoric acid, but this phosphorus is insoluble in water. Practically no rock phosphate, as such, is used as a fertiliser in this country, except some quantity in southern India. In other countries, finely pulverised phosphate rock has been found to give satisfactory results in soils which are very deficient in phosphorous and are acidic. Adequate rainfall and a long growing period of the crop enhance the response. All the same, very little rock phosphate is used is directly as a fertiliser even in these countries. Much more of it is used to manufacture superphosphate, the phosphoric acid of which is water-soluble and is in an available form.

Superphosphate

It is the most widely used fertiliser in India. Prepared formerly by treating bones with sulphuric acid, it is now manufactured largely by treating ground phosphate rock with almost an equal quantity by weight of sulphuric acid. This treatment produces a brownish-grey mixture containing monocalcium phosphate and calcium sulphate (gypsum) in practically equal quantities. This fertiliser is manufactured in three grades : single superphosphate containing 16 to 20 per cent phosphoric acid; dicalcium phosphate, 35 to 38 per cent; and triple superphosphate , 44 to 49 per cent. Single superphosphate is the most commonly available grade in the Indian market. Triple superphosphate in the production of which liquid phosphoric acid is used instead of sulphuric acid, contains very little calcium sulphate. Tripple superphosphate is used mostly in the manufacture of concentrated mixed fertilisers.

The phosphoric acid in superphosphate is wholly water-soluble but when applied to the soil, it is immediately converted into insoluble phosphate owing to precipitation as calcium, iron or aluminium phosphate, according as the soil is alkaline or acid. Thus the fertiliser is not leached, but is slowly dissolved in the soil solution.

Fixation losses can be reduced by applying the fertiliser in bands on both sides of the row of seeds at a depth of 10 to 15 cm with a drill. By this means, at least some of the phosphate does not come into direct contact with the soil and thus the available phosphorous is readily released for absorption by the plant roots.

The fertiliser is suitable for all crops and can be applied to all soils. In acid soils, it should properly be used in conjunction with organic manure. It should be applied before or at sowing or transplanting.

Basic-slag

This is a by-product of steel factories. Depending on the phosphorous content of the iron ore, it contains from 6 to 20 per cent of phosphoric acid (P_2O_5). Slag from Indian steel-mills is poor in P_2O_5 and is not used as a fertiliser. The high-grade European slag containing 15 to 18 per cent P_2O_5 is a popular phosphatic fertiliser in central Europe. It is not as soluble as superphosphate, but unlike the latter, it is alkaline in reaction and good for acid soils. For effective use, it must be pulverised before application.

Bone-meal

The use of raw bones as a manure for fruit trees is an age-old practice in this country. Burying the skeleton of an animal under a fruit-tree is known to benefit its growth and bearing. Large quantities of bones used to be exported until a few years ago. Exports have since declined and bone-meal (i. e. ground bone) is now a widely used phosphate fertiliser. It is available in two forms : (i) raw bone-meal, and (ii) steamed bone-meal. The steaming of bones under pressure removes fats, greases, nitrogen and glue-making substances. Thus, while raw bone-meal contains about 4 per cent slow-acting organic nitrogen and 20 to 25 per cent insoluble phosphoric acid. Steamed bones are more brittle and can be readily ground. This is an advantage, as the rate of availability of phosphoric acid. Steamed bones are more brittle and can be readily ground. This is an advantage, as the rate of availability of phosphoric acid in bones depends largely on their degree of pulverisation. Bone-meal contains only 1 to 2 per cent nitrogen but 25 to 30 per cent phosphoric acid. Steamed bones are more brittle and can be readily ground. This is an advantage, as the rate of availability of phosphoric acid. In bones depends largely on their degree of pulverisation. Bone-meal, having particles not larger than 3/32 inch, considered suitable for use as a fertiliser, but the more finely powedered it is, the quicker its P_2O_5 becomes available in the soil.

Potassic Fertilisers

Most of the Indian fertilisers soils contain a sufficient amount of potash. Potassic fertilisers should, therefore, be applied only to such soils as are definitely known to be deficient in potash or to those which respond to their application, such as sandy soils. They can also be applied to certain crops, such as tobacco, potato, onion, tomato and fruit-trees, to improve the quality and appearance of their produce.

Potassic fertilisers in common use are : (i) muriate of potash (potassium chloride), and (ii) sulphate of potash (potassium sulphate). These salts are important constituents of the waters of oceans and inland seas and of saline deposits derived thereform. The largest known deposits of these salts are at Stassfurt in Germany, in the Caspian Sea region in Russia, in the Dead Sea in Palestine and at some places in California, New Mexico, France and Spain.

Muriate of Potash

It is a gray crystalline material containing 50 to 63 per cent potash (K_2O), the whole of which is readily available. Though highly soluble in water, it is not lost from the soil, as it is absorbed on the colloidal surfaces. It can be applied at sowing time or before sowing.

Sulphate of Potash

It is made by treating potassium chloride with magnesium sulphate and is, therefore, more costly. It contains 48 to 52 per cent K_2O. It dissolves readily in water and becomes available to the crop almost immediately. It can be applied at any time up to sowing, but should not be drilled with the seed. It is considered better than muriate of potash for crops, such as tobacco, chillies, potato and fruit-tree, where quality is of prime importance.

Other sources

Wood ashes, cattle-dung ash, leaf-mould, tobacco stems and water hyacinth are the available indigenous sources of potash. Unleached wood ash contains 5 to 6 per cent of potash in the form of potassium carbonate (which is alkaline), 1 to 2 per cent phosphoric acid and 25 to 30 per cent lime (Cao). Both potassium carbonate and lime in the wood ashes counteract the acidity in the soil. Groundnut shell, paddy husk and bagasse ashes are available near decorticating factories and rice and sugar-mills. These also contain a fair amount of potash and some phosphoric acid. Ground tobacco stems contain 2 to 3 per cent of nitrogen and 6 to 10 per cent of potash in quickly available forms. Hyacinth abounds as a weed in fresh-water ponds in Bengal, Assam, Tripura and Malabar in Kerala. When dry, it contains 1 per cent nitrogen, 4 per cent potassium and a small quantity of phosphorous.

Soil amendments

Lime is generally used for correcting soil acidity, for improving the physical condition of the soil and encouraging bacterial activity. Similarly, gypsum is used for reclaiming 'alkali' soils or land from the sea and improving the structure of heavy black clay soils. Hence, these are called soil amendments.

Compound Fertilisers

These fertilisers are multiple nutrient materials, supplying two or three plant nutrients simultaneously. When both nitrogen and phosphorous are deficient in a soil, a compound fertiliser, e. g. ammophos, can be used. It contains 16 per cent N and 20 per cent P_2O_5. Its use does away with the necessity of purchasing two different fertilisers and mixing them in correct proportion before use.

Mixed Fertilisers

Compound fertilisers contain plant food elements in fixed proportions and are, therefore, not always best adapted to different kinds of soils. Accordingly, the needs of different soils can generally be met most economically by the use of fertiliser mixtures containing two or more materials in suitable proportions. Mixtures usually meet nutrient deficiences in a more balanced manner and require less labour to apply than straight fertilisers used separately. Mixtures containing all the three principal used separately. Mixtures containing all the three principal nutrients (N, P and K) are termed complete fertilisers.

Some manufacturers prepare special mixtures for different crops, such as wheat, sugarcane, paddy, potatoes, tobacco, fruit-trees and vegetables. Trade names like Ammophos, Niciphos and Nitrochalk are employed for other proprietary products, In foreign countries, even insecticides, fungicides and weed-killers, such as DDT, BHC and mercury or copper salts and 2, 4-D are sometimes incorporated into fertilisers mixtures.

When a required mixture is not available of the cost of the mixture sold in the market is high in relation to the cost of its individual components, it may be made at home by mixing the constituent fertilisers in correct proportion. The preparation of mixed fertilisers requires a good knowledge of the properties and mutual reaction of the component straight fertilisers under different climatic and storage conditions. Therefore in preparing such mixtures at the farm or at home, care should be taken to avoid the uneven mixing of incompatible fertilisers, or the mixing which leads to a loss of some of the fertilising nutrients in the form of gas, converts soluble nutrients into insoluble ones or induces caking. It is unwise to mix the following:

1. Ammonium sulphate, ammonium chloride, other ammoniacal fertilisers and nitrogenous organic manures with lime.
2. Sodium nitrate or potassium nitrate with superphosphate.
3. Nitrochalk with superphosphate or lime.
4. Ammonium sulphate-nitrate with lime.
5. Urea with superphosphate.
2. Superphosphate with lime or calcium carbonate or wood ashes.

Ammonium nitrate is an explosive chemical and, therefore, dangerous should be mixing at home. Mixtures containing other nitrates should be made in only such quantities as are to be used immediately, because they

Biological tests involve the growth of seedlings or of lower forms of plants, such as fungi and bacteria, under specified conditions and the study of their relative growth or the content of needed nutrients. A periodic testing of plant tissues for nitrates and other nutrients indicates the changing needs of crops for different food elements. But these are slow and costly processes and hence not always practicable.

The chemical analysis of soils or of plants growing on them constituents the modern method of determining the fertility status of a soil. Such analysis give information on the relative abundance or scarcity of the different nutrients required by crops from the soil, but they give no indication regarding the exact quantity, of fertiliser that may be applied to make good the deficiency. The dependability of this, method can, however, be increased a great deal by co-ordinating its results with those obtained from field experiments.

Facilities for rapid soil-testing have been made available in almost all states and can be availed of. At the same time, a very large number of fertiliser experiments with different crops on a variety of soils are conducted annually in all parts of the country under a comprehensive scheme. The results of these field experiments, when calibrated against those of rapid soil tests, will make the latter purely dependable. This will then be a valuable aid in the hands of extension workers in furnishing advice to farmers regarding fertiliser practices.

It is also sometimes possible to obtain a clue to the nutrient deficiences of soil with the help of deficiency symptoms in plants, as described already. However, a correct diagnosis of deficiency symptoms needs extensive experience. Furthermore, such symptoms in plants appear long after the actual occurrence of nutrient deficiency in the soil. Therefore, such soil deficiencies must be diagnosed and remedied much earlier by adopting other means.

Methods of Applying Fertilisers

In a vast country, such as India, possessing a wide variety of climatic and soil conditions, no definite quantity of any fertiliser can be prescribed as an optimum dose even for one and the same crop for all regions. Each state in the country has conducted fertiliser investigations for the last 50 years and accumulated information on the manurial requirements of principal crops under local conditions. Therefore, for specific fertiliser practices in relation to particular crops and soils, the state agricultural department should always be consulted.

Considerations governing the use of fertilisers. It may be stressed that in order to secure the maximum response to fertilisers, the crop should be irrigated immediately, following their application and at suitable intervals thereafter. The response of a crop under arid and semi-arid conditions is usually uncertain and relatively small. It is, therefore, advantageous to restrict the use of chemical fertilisers mainly to irrigated lands and areas of assured rainfall.

It must also be borne in mind that the maximum profit from the use of fertilisers depends on many factors, such as the nature of soil, the kinds of crop grown, the climate (in relation to soil, the kinds of crops growth), the prices of fertilisers, the market price of the agricultural produce and so no. All these factors must be given due consideration to secure the most economic results.

Changes in one or more of them are bound to influence the economic results. Changes in one or more of them are bound to influence the economics of fertiliser use. The use of fertilisers is also subject to the 'law of diminishing returns'. This means that the rate of increase in the crop yield decreases after a certain point is reached regarding the quantity of fertiliser used, and consequently the value of the additional yield finally becomes less than the cost of the fertiliser. Usually, the smaller applications of fertilisers produce greater percentage increasé of yield than larger applications.

An adequate supply of nutrients is only one of the factors that determine crop yield and that the application of fertilisers is not the sole means of making good the nutrient deficiencies in soils and plants. Equal attention must be paid to the other soil and crop-management practices to ensure good tilt, proper drainage, the required soil reaction, soil conversion, good land use, a suitable crop rotation, adequate organic matter in the soil and satisfactory soil microorganism activity. Each one of these plays a vital role in determining the eventual production. The neglecting of one or more or these factors leads to a reduction in yield and created the need for still heavier manuring. Finally, manuring should not only be balanced in itself, but should also be designed to supplement the good effects of proper land use and beneficial soil management.

Index